Truth Wars

Truth Wars

The Politics of Climate Change, Military Intervention and Financial Crisis

Peter Lee
Principal Lecturer in Ethics and Political Theory,
University of Portsmouth, UK

First published 2015 by
PALGRAVE MACMILLAN

Palgrave Macmillan in the UK is an imprint of Macmillan Publishers Limited, registered in England, company number 785998, of Houndmills, Basingstoke, Hampshire RG21 6XS.

Palgrave Macmillan in the US is a division of St Martin's Press LLC, 175 Fifth Avenue, New York, NY 10010.

Palgrave Macmillan is the global academic imprint of the above companies and has companies and representatives throughout the world.

Palgrave® and Macmillan® are registered trademarks in the United States, the United Kingdom, Europe and other countries.

ISBN: 978–1–137–29847–8 Hardback
ISBN: 978–1–137–29848–5 Paperback

This book is printed on paper suitable for recycling and made from fully managed and sustained forest sources. Logging, pulping and manufacturing processes are expected to conform to the environmental regulations of the country of origin.

A catalogue record for this book is available from the British Library.

A catalog record for this book is available from the Library of Congress.

For Lorna, Samantha and Fiona

Certainty is a false friend in the quest for truth.

Contents

Preface

One of the most enjoyable aspects of researching and writing this book has been the reaction of friends, colleagues or even complete strangers when I have mentioned that it explores the relationship between politics and truth. There has been a remarkably similar response on almost every occasion: nervous laughter followed by, 'Well that will be a short book!' A common perception seems to be that politics – in particular, the governing done by professional politicians – is characterised by dishonesty, deception and untruths. Given the frequently self-interested antics of elected representatives it is easy to see why such attitudes prevail. 'They just tell you what they think you want to hear, so you will do what they want,' is another accusation that is thrown at politicians, as though everyone does not play that particular game on a daily basis: telling others what we think they want to hear in order to shape their behaviour in some way.

Yet politics is much broader, more inclusive, than the actions carried out by governments and professional politicians. Politics permeates every aspect of life, from its most public to its most private and intimate aspects. And the more I paid attention, I mean *really* paid attention, the more it became clear that truth – or at least claims and counter-claims concerning truth – pervades every aspect of politics, from the personal to the global. I will go further and make a truth claim of my own (inevitably contestable), that the essence of politics is a perpetual, overlapping series of truth wars, fought for the purpose of shaping individual conduct, attitudes and identities. These truth wars take many forms and are prompted by competing claims across multiple domains, including political ideology, religious belief and practice, scientific discovery, the application of logic, philosophical ideals, moral frameworks, economic theory, product advertising, cultural norms and countless more.

The scale of global crises only intensifies the ferocity of the truth wars at their core because of the potentially vast human, environmental and economic costs, and the competing interests, involved. The three crises chosen as the focus of this book – climate change, military intervention and global financial crisis – highlight different aspects of the way truth is created, presented, deployed and disputed in the course of shaping, and encouraging us to self-shape, our lives in particular ways. Complicating matters further is the way that language is used in the process. Language does not merely describe the social world in which we live, it *creates* that world. Similarly, language does not simply describe truth, it is used to create and shape truth in specific contexts. By exploring three twenty-first century truth wars, I do not set out to show you, the reader, what the truth *is* about any of them. Rather, I try to demonstrate how, and the extent to which, truth is constructed, sustained and disputed in the process of shaping the way individuals understand themselves, major political crises, and the relationship between the two.

Acknowledgements

Many of the ideas in this book were developed over time in lectures and conference papers I have delivered and in the cut and thrust of public debate. I do not know the names of many of the interlocutors who have challenged and shaped my thinking in recent years but I thank them all the same. As the manuscript has taken shape, friends, colleagues and knowledgeable individuals previously unknown to me have been generous with their time in offering detailed comments and critique. In particular I would like to thank Bettina Renz, Andrew Mumford, Alan Tait, Dominic Lawson, Adam Sutch, Jason Sannegadu, Rob Spalton and Kimberly Hutchings. They have helped to smooth the roughest edges off the chapters to follow but bear no responsibility for my arguments or remaining errors, either in whole or in part.

Scripture references are taken from the Holy Bible: New International Version, Copyright © 1973, 1978, 1984 by International Bible Society. Anglicisation Copyright © 1979, 1984, 1989. Used by permission of Hodder and Stoughton Limited.

Finally, my greatest thanks go to my wife, Lorna, and our daughters, Samantha and Fiona, who give me more love, encouragement and support than anyone has the right to hope for.

Introduction

We live in an age of crisis. There are many types of crisis, of course, but the three that serve as the focus for this book – climate change, military intervention and financial crisis – are widely claimed to be global in scale and potentially apocalyptic in severity. These crises all have one thing in common: they each provide political leaders with the incentive and justification to increasingly govern the lives of millions, even billions, of people through the enactment of policy and the allocation, or withdrawal, of resources. They add a new dimension to Harold Lasswell's famous aphorism that politics is about who gets what, when and how,[1] because such decisions are based on specific truth claims and the policy priorities that emerge from them. For example, ideologies like capitalism and socialism each make claims about what is 'true', which then shapes behaviour and dictates what counts as ethical practice. Complicating matters further, powerful interest groups – banks, multinational corporations, environmental organisations, and others – have increasingly efficient lobbying networks that, to varying degrees, subvert the democratic process.

In response to these crises, the language of war has been increasingly deployed across a whole spectrum of ecological, social and economic problems: war on terror; war on warming; war on want; war on bankers' bonuses; war on drugs; war on waste; war on genocidal leaders – with each new declared 'war' subverting Carl von Clausewitz's formulation that war is merely politics by other means and giving credence to Michel Foucault's inversion of that idea: that 'politics is the continuation of war by other means'.[2] 'Truth' is the terrain over which this war of global politics is being waged. For example, when a shockwave tore through the global financial system in 2008 to cause the most widespread economic crisis in history, it was initially met with denial, bemusement, more denial and incredulity, quickly followed by panic, despair, gradual

1

acceptance, plans for avoiding a repeat, and a universal desire to apportion blame. Why did no one see it coming? Where did all the money go? Who is responsible?

In the years since, countless studies have sought to understand why there emerged a sudden, surprise shortage of money and an abundance of bad debt that could never be repaid. One crude answer is that credit levels in both government and private finances had been ballooning for several years, fuelled by increasingly novel financial products and practices. Effectively, an increasingly precarious debt skyscraper was being built from loans being made upon loans, with insurance conglomerates wagering that the building would keep rising. It didn't.

Contrary to the expectations or perhaps naïve assumptions of vast numbers of individuals whose jobs, homes and futures relied upon the debt skyscraper reaching towards infinity, the edifice was not built upon foundations of gold or silver or any other substance of 'real' value. The tower of debt was built upon nothing more than illusory truth claims, a central pillar of the illusion being that expert economists and financiers knew what they were doing and were in control of events. From the political arena to the private household, belief that those good times would continue to roll were at least as strong as in the financial markets. Everyone wanted it to be true, and the overwhelming majority ignored the sceptical few who dared to point out that it could not be.

So what is 'the truth' about the state of the US and UK economies from 2006 to 2008? On this issue, as with the others addressed in this book, I set aside the notion that there is a universal, definitive, underlying truth somewhere just waiting to be discovered. When it comes to shaping truth and claiming it for themselves, governments and powerful national and international institutions – guided by decision-making individuals in positions where they can wield great power – operate within their own reality, motivated by a combination of self-interest, national interest (sometimes), ideology, and a desire to maintain the power and positions they hold. When Paris Welch, a California mortgage broker, wrote to US financial regulators in January 2006 with her concerns about the dangers and potential consequences of risky lending practices, she cautioned: 'Expect fallout, expect foreclosures, expect horror stories'.[3] She was ignored and her concerns swept aside.

It is easy to look back with the benefit of hindsight and an internet search engine and compile a catalogue of prophetic voices that pointed to the disaster. History shows, at least on a crude calculation, that the assessments of Welch and many others can be viewed retrospectively as true, while those of some of the most powerful figures in the global economy can be seen as untrue. So why were they rejected at the time? Similar questions were being asked and truth claims disputed in other major crises: from the need for military intervention to prevent huge loss of life at the hands of dictators, to the need to levy global carbon taxes as a means of reducing global warming. Consequently, this book explores the nature of truth wars in the international political arena, and the relevance of those claims and counter-claims in the shaping of human identity, attitudes and conduct.

Truth, truth claims and regimes of truth

Truth wars do not exist for their own sake but play a crucial role in governing behaviour: first, through the direct exercise of power; and second – more subtly and indirectly – by prompting or inciting individuals to conduct themselves and shape their identities in relation to truth claims. In addition, a common element that connects truth claims, the exercise of power, and personal choice in the forming of one's own identity and the shaping of one's actions is ethics: doing the right thing – and resisting the wrong thing – in a complex world.

Perhaps the greatest difficulty faced by anyone who wants to analyse, argue, dispute or otherwise engage with truth as a subject matter is the timeless challenge of understanding what we might mean by the word. Every news bulletin and every newspaper contains multiple claims to the truth, usually spread over a number of domains, including: truth as a somehow indisputable fact or logical progression; truth as a verified scientific finding; ideological truth, which *de facto* rejects and subverts its opponents; observational truth ('I saw it, so it must be true' – a favoured approach in alien abduction and other conspiracy theories); faith-based religious truth, every one of which identifies the beliefs of its opponents or alternatives as heretical; aspirational truth ('I hope it is true'); truth as a philosophical entity with an existence of its own that is merely waiting to be uncovered; semantic truth (sentences with an internal logic). The list of possibilities is endless.

While this book investigates the notion and application of truth in the political arena – domestically and internationally – I do not set out to identify and lay claim to *the* truth. I reject the possibility that there is one overarching political truth simply waiting to be revealed: truth is always contested, and on multiple grounds. Rather, it will explore some of the ways in which truth *claims* emerge, some of which gain such a degree of acceptance and authority in particular societies that they come to acquire the status of what Michel Foucault described as *regimes* of truth.[4]

A regime of truth is considered here to have gained social, institutional, and governmental – perhaps global – dominance to the extent that for its supporters its validity is beyond question: democracy itself would be one such example. For opponents, however, it will be viewed as illegitimate and subjected to critique, rejection, modification and subversion.

The purpose, ultimately, of the truth wars is – to coin a phrase popularised by George W. Bush (who said he wanted it) and Tony Blair (who said he didn't want it but actually did) in relation to Iraq – regime change. That is, supporting or changing a particular regime of truth to favour one side or the other. For example, Greenpeace and the Intergovernmental Panel on Climate Change will contest climate change truth claims with major oil companies. The chapters to follow will explore how multiple and overlapping conceptions of truth are deployed in the political arena: from scientific claims and counter-claims in relation to climate change, and the supposed 'real' reasons why governments choose to intervene, or not to intervene, militarily in the affairs of other states, to the 'truth' about the global financial crisis.

But what is it about statements that might make them true or otherwise? Is it where propositions correspond with things known as facts? Is it where every part of an utterance contains both an internal and external logic? Or does it depend on the context in which truth is claimed or disputed? By that I mean: is it reasonable to speak of scientific truths separately from social scientific truths or more esoteric philosophical or theological conceptions of what truth is, or even to judge such claims in the same way? It only takes a few cursory questions to reveal that any search for an all-encompassing, meaningful definition and application of truth in the political realm is futile in a book of this size. It is probably futile as a life's work, given that philosophers have sought out the meaning of truth since the first human ancestor responded to an accusation with a variation on the words: 'Honest, I didn't do it – that's the truth.' Debate started in earnest when the accuser responded: 'Yes he did – I saw him.'

Different schools of thought have attached particular criteria to the ability to refer to something as true.[5] For some, understanding is a necessary component of truth. If we do not understand aspects of a particular debate then it is not possible to ascertain which of the disputing parties is promoting the truth. And even if we understand all the issues it might be the case that each of the protagonists clings partially, but only partially, to an element of the truth – according to how we may wish to judge or define it. Alternatively, in the fifth century Augustine claimed that the Christian's understanding (of religious *truth*) comes through faith: with faith setting the parameters for what could count as true. However, such an approach could hardly be more different to modern understandings of truth and knowledge that emerge from scientific processes and their emphasis on proof and the ability to replicate findings.

A considerable number of people hold global warming or climate change to be true, or conversely, untrue, without having read a single scientific paper on the subject. Such people might be more accurately said to be exercising faith or trust in those who have read – and written – such scientific papers and who therefore may be in a better position to declare truth or falsehood, at least in scientific terms. Perhaps for some, holding a position on such a debate says more about the world *as they would ideally like it to be* than about the reality they face. In other words, they may be more motivated by truth that takes the form of political ideology, faith or hope than with truth as something that can be proven: a disconcerting thought in light of the popularisation of suicide bombings as a military tactic among some political groups. Consider two ways in which approaches to truth from the past century can be simultaneously helpful and unhelpful when exploring the truth wars that characterise twenty-first century political crises – Truth in language and Truth, identity and behaviour.

Truth in language

In the mid-twentieth century Alfred Tarski set out a definition for truth in formalised language use. He noted that 'A sentence is true if it designates an existing state of affairs'.[6] He began to illustrate what he meant with what is probably his most famous formulation, an analysis of the conditions in which the sentence 'snow is white' can be deemed either true or false.[7] Perhaps unsurprisingly, he concluded that 'The sentence 'snow is white' is true if and only if, snow is white'.[8] The apparent simplicity of this finding – the less charitable might say statement of the blindingly obvious – belies the complexity of the mathematically-influenced analysis that he conducted. However, even stopping at this point and applying this idea to global warming and climate change we see two different truth conditions emerge.

Consider the sentence: 'The globe is warming.' Using Tarski's logic this statement is true if, and only if, the globe is getting hotter. This statement contains the drawback that there are two conditions under which it can be said to be untrue: if global temperatures remain constant over some specified period or if global temperatures fall. Contained within the phrase 'global warming' is the logical possibility of its own downfall as a truth claim. Alternatively, reflect on the sentence: 'The global climate is changing.' Again, using Tarski's logic, the sentence is true if, and only if, the global climate changes. Given the complexity of the Earth's climate and the extremes of temperatures and weather patterns that have been achieved over millions of years, it is difficult to imagine any time or circumstance where the sentence, 'the global climate is changing' can be said to be anything but true. And that is before the influence of political ideology is brought to bear on the question. Logic alone dictates that, if an individual or group were seeking to promote a particular view of the Earth and its climate, it makes sense to choose a descriptor that must be true in every circumstance as opposed to a descriptor that offers the possibility of falsification or failure. It is the climate equivalent of hedge fund management, where profits can be made both when share prices rise and when they fall. Unfortunately for those who seek to understand how the world works and why people make particular claims in relation to the truth, humans and the societies they form interact in ways that are as complex and unpredictable as weather patterns. Consequently, Tarski's sentence, 'Snow is white,' offers only limited practical potential for understanding truth claims in relation to human conduct and political decision-making.

Take another snow-themed sentence, which this time introduces a behavioural element: 'Politicians ski on white snow'. There are four potential variables in this sentence – politicians, skiing, the colour white, and snow – thereby making it more complex than Tarski's original. It does not take long to identify some of the difficulties in seeking out truth once human activities, choices and judgements are involved (it is not a coincidence that I am using the classification of 'politicians' to illustrate my argument here), especially if said politicians have control of state institutions and apparatus with which to reinforce their arguments.

What happens if the snow is slightly muddy? The politicians might want to ski because they see it as a means of gaining popular support or as acting in the best interests of the people they represent, but at the same time they do not want to render the sentence, 'Politicians ski on white snow,' untrue. One possibility is that they redefine the meaning of 'white'. After all, house paints come in a variety of 'whites' which can contain traces of green, pink, yellow, or blue, and so on, so it makes sense to extend the range of shades that can be called white. In a similar vein, strange weather patterns might result in the melting and re-freezing of the snow surface so that it is now ice that the politicians are skiing on. Naturally, if the meaning of 'white' can be extended to include shades that were not previously classed as white, then – some might argue – there is no harm in extending the meaning of 'snow' to include 'ice'. Finally, several of the politicians might decide that for the sake of safety they would prefer to sledge down the snowy hills rather than ski, so once more a meaning is extended to make sledging the new skiing: for the best of intentions of course.

Some time later, a citizen of the politicians' homeland returns to witness a peculiar sight. She sees politicians sledging on brown ice, while still insisting that they are upholding the 'truth' that 'Politicians ski on white snow'. While Tarski's utterance that snow is white might offer a means of identifying some objective truth in the narrow, logical context of sentences, that truth – even in the simplistic example I have used here – is unlikely to survive engagement with the idiosyncrasies, desires and ambitions of human beings: whether their stated motivations are selfish or selfless. If 'truth' in the murky world of politics was generally applicable, easily recognisable and simple to grasp, the problems of military intervention, economic recession and climate change would have been agreed by now and solutions set in motion, all with universal support. Unfortunately, that idealised approach does not resonate with most people's experience of political reality. Consequently, I will not attempt to add another voice to those who seek, and often claim, intimate familiarity with some objective 'truth' about the politics of global crises. Rather, it is a study of how, in the exercise of power, truth is established and deployed at a subjective level to shape human behaviour and identity in the pursuit of political ends.

Truth, identity and behaviour

So where does that leave the search for truth in relation to climate change, military intervention, and financial crisis? Any philosophical enquiry concerning truth eventually faces the question: what is truth? I have already rejected the assumption that truth has some objective existence of its own, disconnected from the corrupting world and waiting to be discovered by the earnest seeker with the pure heart. Instead I view truth as something that is produced within complex relations between the individuals who wield political power, those who manage vast economic resources, the people who control the institutions and mechanisms that validate or invalidate scientific or other knowledge claims, and those who are subject to that power.[9]

The chapters that follow will explore the politics of truth in relation to climate change, military intervention, and financial crisis, looking not for some definitive, objective truth or falsehood but for the ways in which opposing protagonists stake their claims and make their arguments: shaping individual conduct and identity in the process. For example, rather than adopting a pro or anti-climate change stance and attempting to assess the scientific argument in an imagined detached, objectivist way, I will examine how the individuals are characterised – and self-characterised – within the debate as either 'denier', 'alarmist' or neutral: with mechanisms of state and scientific institutions used to strengthen particular arguments. In so doing I will examine the ways in which the political debates draw upon and deploy established aspects of scientific, social scientific and other discourses, as governments and other powerful groups use truth claims to shape individual lives, either directly or indirectly. Direct shaping of identity and behaviour can take the form of policy-making that codifies expected conduct. In totalitarian regimes propaganda is used aggressively against populations for similar ends. Indirect shaping of individuals in liberal democratic contexts is perhaps more subtle and interesting, as people are encouraged or incited to form their own identities, beliefs and behaviour, usually in ethical terms, as a work of self upon self: 'If you are a good citizen you will do the right thing and support this or that political policy…recycle household waste…pay higher "green" taxes… support military intervention,' and so on.

Many books and articles on the themes to be addressed here attempt to make *objective* analyses of the many problems they present: reaching for the big picture that explains the extent of the problem and seeks to convert readers to, or confirm them in, a particular world view. In contrast, this book will concentrate on the *subjective*, attempting to understand how, where and perhaps why the individual is the focus of particular arguments at particular times. In a world where science frequently holds a dominant claim to truth based on the dispassionate and objective pursuit of knowledge, the subjective is frequently, easily and erroneously dismissed. Knowledge and power are the key weapons in the truth wars that shape our day-to-day attitudes, beliefs and actions, and they are deployed simultaneously in contrasting and effective ways. Means of control include – but are not limited to – law-making, law enforcement, security apparatuses, education and officially-sponsored surveillance.

From local authority by-laws concerning the fly-tipping of waste, to the collection of taxes by the government, all underpinned by the threat of prosecution and incarceration, there are countless ways in which our lives and choices are acted upon. For example, the US Patriot Act[10] has both impassioned supporters and equally vociferous critics. Supporters might prioritise the survival and well-being of the Republic at the expense of individual freedoms, while critics might be willing to live with greater threats so long as rights and liberties are not constrained by the state. Introduced in response to the Al-Qaeda terrorist attacks of 11 September 2001, the Patriot Act gives sweeping investigatory powers to the US security services. Much has been written already about the extent to which it encroaches on the freedoms granted by the US Constitution to the American people – ironically, in the name of defending the Constitution and the American people. Of interest to the approach taken in this book is not some attempted objective cost/benefit evaluation of the Act itself, but rather how it came to be so widely accepted. Its very name – Patriot Act – implies a descriptor of its opponents: non-patriots. To support or reject this Act is for individuals to constitute themselves as particular kinds of people: patriotic, dutiful, lovers of America, or non-patriotic, anti-American, Al-Qaeda sympathisers. This example is worthy of a study all of its own but hopefully the point has been made that the subjective dimension of politics, though frequently overlooked, deserves to be explored further.

In addition, we are encouraged to work on ourselves – our characters, our bodies, our psychological demons, our confidence, our religious faith or atheism – in numerous ways. Populations are bombarded with visual imagery and other messages that promote a particular view of beauty, physical shape and lifestyle. Cosmetics companies will use models of different ethnic origin to promote their products in different parts of the world. It is no coincidence that lifestyle magazines are financed, through advertising, by companies who want to make money by selling everything from cosmetic surgery to diet books, and from perfume to anti-wrinkle cream to help us achieve our individual aspirations. We are first sold the aspiration, the dream; then we are sold the products that will supposedly help to achieve it.

More subtle are the ways in which we are encouraged to work on other aspects of our identity and conduct: anti-smoking campaigns, alcohol awareness messages, recycling practices, eating five portions of fruit and vegetables per day, to name but a few. So much so that it is easy to lose sight of when powerful bodies – including national and local government, multinational corporations such as McDonalds, and large pressure groups like Greenpeace – are trying to manipulate our attitudes and conduct. The internet adds another dimension to the ways in which our lives are shaped and our purchasing influenced. Pop-up advertising is targeted at individuals based on previous browsing and purchasing habits: personal information that is stored by companies for the express purpose of marketing. There is a common misperception that when users go to the Google or other search page that they are somehow Google's customers. Wrong. Google's customers are the commercial entities that pay to target their marketing through the Google website or via the information that Google harvests. The person using the Google search engine to locate information is the *product* that Google sells – an uncomfortable truth and one that is not widely publicised. The whole notion of targeted advertising is premised on getting an individual to change his or her behaviour by buying a product that will somehow enrich their lives, promote a particular aspirational lifestyle or, in the case of charity appeals, somehow make the world a better place and the giver a better person. From war to climate change to financial products to advertising, the shaping of behaviour and attitudes rests firmly on presenting and reinforcing particular truth discourses and encouraging people to accept those truth claims and change their lives accordingly.

In considering the ways that individuals are shaped by the truth wars waged around political crises, more questions emerge: On what basis do I discover or question the 'truth' about any given situation? Do I rely on personal conscience, which in turn may have been shaped by particular social or religious assumptions? Or is it a question of rigorous intellectual engagement, weighing up the pros and cons of every argument (assuming I have sufficient, and sufficiently accurate, information with which to work)? For the Google-Wiki generation perhaps it is just a case of reaching for the nearest internet connection and accepting the truth and accuracy of what I find. Also, is there a certain way that I am expected to behave if I want to be seen, and see myself, as ethical: by doing 'the right thing' that so many politicians in particular encourage their listeners to do? The 'right thing' in this instance is usually to follow the politician's chosen path, which leads back to the need to make judgements about truth. Where in the past adherents of particular religious or ideological truths identified themselves with public declarations or confessions, today's seeker after truth is more likely to set out their position on their Facebook page, Twitter account or in a 'selfie' portrait that flashes up on Instagram. The truth is what we broadcast, 'like' or recommend to others.

Then comes the challenge of changing myself to be someone better, becoming an increasingly ethical being who takes seriously his responsibilities in the world and to the world. Where I accept a particular truth claim in the truth wars (that is, take sides), how does it change me, my attitudes and my sense of self and purpose? And finally, what is the point of all of this – what am I trying to achieve? A better world for future generations; survival; justice; eternity in heaven; a martyr's welcome in Paradise? Someone who is motivated by the dream of a martyr's death is likely to behave quite differently to someone who wants to safeguard the ecological system by living in harmony with nature and fellow human beings. Yet both may see themselves as motivated by ethical concerns.

In light of these questions the struggle for truth in political crises will be explored as follows. The first section addresses the truth wars at the heart of climate change. At this point it becomes obligatory to set out my own position on climate change, so that some readers – not all I hope – can decide whether or not it is worth reading on. That is how it is with climate change; positions are entrenched and we like to know how much someone agrees or disagrees with us. So, for what it is worth, I am convinced that climate change exists. I am also convinced that it has always existed. I am further convinced that human behaviour has contributed to climate change, especially over the last century or so. That is where things get interesting because the precise extent of the human contribution to climate change has, for me, yet to be established: I reject any claim that there is no (zero per cent) human contribution and I reject the claim that it is all (100 per cent) down to human behaviour. With personal confession out of the way, for those who are still reading, Chapter 1 will examine the politics of climate science, which has prompted particular ethical responses as global warming shifted very rapidly from being an obscure area of esoteric scientific enquiry in the 1970s to being one of the highest profile social and political issues in domestic and global policy making. Chapter 2 moves on to the science of climate politics and the means by which climate change activists and advocates have, over the past two decades, sought to change the behaviour of millions, even billions, of people, and how opponents – so-called deniers and contrarians – have fought back with increasingly marginalised claims to truth. I explore the means by which each side in the climate change debate constructs its 'truth', not simply by an objective analysis of scientific evidence – though it plays an important role – but also through the selective reading of the philosophy of science, and the use of political and academic structures as a means of shaping individual identity and behaviour. The third chapter examines the ways in which individual identity emerges in environmental discourse – as a sustainer or a harmer of the natural world. In 1972 the Club of Rome articulated a highly subjective position on the problem, stating: 'The real enemy then is humanity itself'.[11] The ideological underpinnings of the global response to climate change discourse are investigated as part of the truth wars that will shape climate politics and policy for generations to come.

The second section of the book turns to military intervention discourse. Where the vast majority of the literature on intervention is framed in terms of either the moral right to intervene in another state's affairs or the legal right to violate international borders to do so, these chapters maintain a focus on the individual and examine how different aspects of identity shape debate. Chapter 4 challenges the assumption that opposing tyranny is an inherently good and desirable action to take in the international system, and that it will inevitably lead to freedom and democracy. The fifth chapter looks at the role of gender in the justification and execution of recent military interventions in Afghanistan, Iraq and Libya, highlighting how gender stereotypes have proved to be remarkably enduring and exploited for political ends. Chapter 6 considers a recent technical innovation – the weaponised drone – and explores the extent to which it changes the terms of military intervention. While the term 'drone' connotes autonomy and a lack of humanity, and public discourse is dominated by visions of detached killing machines, humans are heavily involved in, and deeply affected by, every aspect of their use: from the drone operators on one extreme to the targets on the other.

The final section shifts the focus of discussion to the financial crisis that hit the global economy in 2007–2008, with particular emphasis on events in the US and UK. Chapter 7 sets the crisis in context, outlining key factors that contributed to the collapse in the financial sector, before going on to show how reactions became increasingly personalised. Two specific groups are examined, political leaders and subprime borrowers: showing how the identities, attitudes and behaviours of those involved were shaped in the global game of blame and counter-blame. Chapter 8 continues the theme of blame attribution, examining how bankers have been portrayed, and have portrayed themselves, both before and during the financial crisis. The identities, conduct, and attitudes of senior bankers will be assessed against the codes that were supposed to constrain their excesses and protect the public, in light of the financial truth wars that have waged over recent years in an attempt to govern greed. Chapter 9 concentrates on one specific geographical area affected by ongoing recession, the political upheaval that has accompanied the struggle to achieve economic growth: Europe. Attempts to stabilise the Euro as a currency and the Eurozone as a credible, unitary political entity are increasingly being framed in personalised terms. Analysis of these events shows how the return of national stereotypes – reliable, frugal, hard-working Germans; tax avoiding Greeks; disorganised, undisciplined Italians; and Spaniards who prefer siesta to work – has been exacerbated by the very political structures that were meant to prevent such divisions in Europe from arising again.

The Epilogue draws together some of the themes that have emerged, summarising key lessons about the relationship between the individual and claims to truth during times of political crisis. Commonalities across the different strands of political discourse are identified, as are divergences and disagreements. Ultimately, the reader will have a greater understanding of why truth wars occur, their ongoing inevitability, and how they are used to shape identity, attitudes and behaviour.

Part I

Politics, Truth and Climate Change

1
Climate, Science and Truth

'Of course I know it's true – scientists have proved it. You're a classic Denier!'

'You don't know it's true because scientists have shown otherwise. Typical hysterical Alarmist!'

It had taken no more than three minutes for what had started off as a discussion about global warming to degenerate into something more akin to a group of eight-year-olds squabbling over the existence of Santa Claus. For some it was a FACT. End of. For others it was made-up mumbo jumbo. For a more thoughtful few it probably was real but they worried that maybe it wasn't. Others believed it probably wasn't true but did not want to look foolish in case they were later shown to be wrong. Everyone had a view on what should, or should not, be done about it and *nobody* was willing to change their position. The entire climate debate in abrasive microcosm: scientific claim and counter-claim, aggressive name calling, and words like 'evidence' and 'truth' wielded like weapons by speakers unqualified to define either.

What had caused such an instant storm of fury in a class of 120 intelligent, highly motivated undergraduate students who were meant to be considering global security challenges? Someone, probably mischievously aware of the conceptual hand grenade he was about to unleash, had put his hand up to interrupt me and asked: Is global warming real? Before I could respond and try to keep my lecture focused on the task at hand, replies to the question came thick and fast from every part of the lecture theatre. Lecturers normally dream of energetic, engaged and passionate audiences, especially when the day-to-day reality is often undisguised apathy nursing a hangover and wondering how little effort can be expended while still gaining the necessary course credits to pursue life's dream. This, however, was more of an intellectual riot than informed, high brow debate.

Intrigued by the unexpected change of direction, I decided to conduct an impromptu and highly unscientific straw poll of the 120 students present. On the question, 'Has science demonstrated the existence of human-caused global warming?' a clear majority – around 80 of the students – supported the view that the world is getting demonstrably warmer and human activity is responsible. Twenty took a different view: that even if global warming is happening it is part of the natural variations in Earth's climate. And a further 20 felt they were not in a position to say. I asked the students how many of them had read a published academic journal (as opposed to newspaper, magazine or internet) article on the subject. Six students put their hands up – each of them had studied at least one undergraduate course module on the environment, global warming or climate change.

The most fascinating aspect of the whole interchange was not the extent to which science had proved the truth of global warming or otherwise. For me, the most interesting question was this: How did so many people who had never read a scientific paper on global warming come to make such strong truth claims on the subject? Almost as interesting was the question: By what means were these particular truth claims about global warming mediated so effectively that dozens of intelligent students had committed themselves to what appeared to be – given their lack of scholarly evidence-gathering on either side – statements of faith.

The students' responses to these questions prompted this book. The most common sources of information – excluding those who had studied courses on ecology, environmentalism or climate change – were television, school and the internet. Almost all of the students had been in primary (junior) school at the time of the Kyoto Protocol in 1997 and most of them could remember doing projects to predict or imagine what the world would look like *when* (not if) it was several degrees hotter, coastlines were flooding and there was not enough food being produced to feed the world's population. That so many students could remember carrying out such activities at the age of 7 or 8 is testament to a combination of the effectiveness of their teachers' efforts and of the impressionability of children at that age. Yet it is barely conceivable that these students, most from disparate parts of the UK and a handful from other continents, all had teachers who had themselves been poring over the scientific literature on global warming. Quite a few students volunteered that their views on global warming were probably fixed at primary school in the late 1990s and early 2000s and, apart from the six who had taken courses on the subject, nobody was aware that there had been no statistically significant rise in global mean temperature since 1997 (at least as of 2013 – a distinct upward trend may have continued by now).[1]

Facts, knowledge and understanding were definitely thin on the ground but opinions about people who took opposing or contradictory views were not. There was generally a far greater engagement with subjective aspects of climate debate – self-identity and the identity of others – than any pretence at an objective weighing up of evidence. Consequently, the question that emerged from my reflections on the classroom encounter is this: How objective (a word whose meaning is highly contested anyway) is the climate science that shapes opinion, identity and behaviour so profoundly in so many observers?

The remainder of the chapter will explore this question in relation to the politics of climate science. It will become apparent that claims to objective understanding by both proponents and opponents of the global warming thesis overstate what is knowable in climate science. And further, that subjective elements – individual identity, values, ideology, even faith – play a much more significant role in the debate than is typically acknowledged in the scientific community and beyond. The first section of the chapter will examine what scientists claim for science and the degree to which scientific method can be said to produce scientific knowledge or truth, especially in the context of global warming and its offshoot, climate change. The second section will look at the personal politics of climate change discourse and how they have come to play such a key role in public debate, usually as a means of avoiding, not reinforcing, demonstrable scientific claims: all in the process of shaping individual attitudes and behaviour with regard to the environment. The final section of the chapter will look at ways in which individuals are encouraged to form their own identities – particularly as ethical beings – in climate debate, alongside the ways in which identity is constituted in relations of power.

Science, climate science and truth

Many people – like the most vocal protagonists in my impromptu global warming debate – *think* they know what science is, probably based on their experience of physics, biology and chemistry in high school. However, even these superficial categories of science are highly distinct, each having its own disciplinary focus and practices. Further, the precise nature of rational scientific explanation has been disputed even by individuals and schools of thought that accept its existence:

from the early positivists who set out to remove metaphysics, values and beliefs from scientific knowledge, to Karl Popper who separated science from non-science or pseudo-science on the basis of whether it was testable – capable of being falsified – or not. For Popper, perception is not to be trusted – only observation, strictly applied, can lead to true knowledge.[2] Complicating matters, according to Paul Feyerabend, philosopher of science and arch-critic of Popper and other rationalist approaches to science, the word 'science' does not correlate to any one particular activity or entity, which makes it difficult to understand what science *is* and therefore what knowledge or 'truth' it might claim to produce.[3]

Since it is beyond the scope of this book to enter into an extended discussion of the philosophy of science, I will summarise the difficulty facing any scientific endeavour by highlighting that the very meaning of 'science' is hotly disputed and constantly challenged, both between different fields and within specific areas of interest. However, in developing an understanding of any subjective element of science we need to start somewhere, if only to show how vulnerable definitions are to being undermined.

The US National Academy of Sciences provides a definition of the term scientist, which it 'applies very broadly and includes all researchers engaged in the pursuit of new knowledge through investigations that apply scientific methods'.[4] In this description the identity of the scientist is based on a specific way of working that rests ultimately on the application of scientific methods. In turn, modern scientific methods have emerged since the Enlightenment and are used in different ways, in the pursuit of different ends, by radically different schools of scientific thought: there is no unified meaning or universally accepted definition. In 1942 Robert Merton codified ideals that underpin what many would associate with 'normal' science: universalism (truth claims tested on universal, unbiased criteria), communalism (where scientific findings are commonly shared), disinterestedness (acting in a way that is not shaped by vested or personal interests), and organised scepticism (findings should be rigorously scrutinised).[5] The word 'normal' will take on increasing significance as the chapter progresses. Thomas Kuhn has extensively critiqued the notion of 'normal science', and it is worth acknowledging one of his criticisms: namely that normalised approaches, by definition, tend not to see or explore major novelties.[6]

Scientific activity spans both theory and practice. Theory prompts hypotheses that can subsequently be validated or invalidated through observation and experimentation: all in pursuit of the idealised goal of objectivity. However, as theory, hypothesis, observation and experimentation are combined in limitless ways in pursuit of scientific knowledge, the subjective input from scientists is frequently ignored, denied or minimised. Combinations of scientific activities must be selected and applied in accepted ways by scientists according to what they are trying to achieve. A scientist will *form* one hypothesis while rejecting an alternative possibility; a suitable method will be *chosen* and an experiment's parameters *defined*. With each choice there is an accepted, unquestioned and usually overlooked subjective dimension to the pursuit of objectivity within the confines of the scientific method, however defined.

One of the difficulties in voicing even the slightest concern about science and the way it works – especially in climate science – is that so many scientists, deliberately or subconsciously, see their specialist field as unique, singularly important and therefore immune to criticism from 'outsiders'. These outsiders can be opponents within a field of study, they might emerge from what might be described as the broader scientific community and, with the least credibility of all, and they might even be non-scientists. Such territoriality is frequently misplaced, especially when it comes to trying to understand the politics of science, and hence the politics of science-based truth claims. It is more accurate, or at least honest, to say that the science of science is disputed, indecisive and temperamental. The happy optimist or convinced practitioner might be content to say that science is science and that it is inherently authoritative but the rest of us would do well to ask in greater detail what is going on when someone says they root their arguments in science. Take a relatively non-controversial example of science at work.

On 4 July 2012 the global scientific community (if there is such a thing) witnessed the announcement of a remarkable success that had been 50 years in the making and which illustrates the pursuit of science-based truth or knowledge at its contentious best. With apologies to particle physicists for whom this subject is their life and passion and who know an interloper when they see one, scientists at CERN, the European Organization for Nuclear Research, stated that they had observed a number of Higgs bosons. The Higgs boson was first hypothesised in 1964 by the theoretical and particle physicist Peter Higgs, as a means of explaining how particles gained mass as they exploded out from the Big Bang at almost the speed of light. As particles passed through a Higgs field – made up of Higgs bosons – they would acquire the mass that would give form to the universe. Problematically, the Higgs boson is so tiny and decays so rapidly after coming into existence that it took several decades, vast expense and a huge consumption of energy by the Large Hadron Collider and its research teams in Switzerland to finally provide a glimpse of it. Note the sequence: Higgs made the hypothesis in 1964 to make his particle theory 'work'; experiments were devised and conducted over a period of almost 50 years; then, finally, the particle was observed and the observation compared with the original theoretical proposal.

Finally, Rolf Heuer, director-general of CERN, announced: 'We have a discovery – we have observed a new particle consistent with a Higgs boson'.[7] Heuer's statement was calm, cautious and qualified, his approach echoed by The Atlas Collaboration in the official publication of the experiment's results: 'The high degree of statistical significance and simultaneous observation in multiple channels and data sets in this search for the SM Higgs boson demonstrate that we have observed a new particle with properties consistent with those of the SM Higgs boson.'[8] Although the initial findings were replicated and confirmed by a second, independent group of researchers, the claims to new knowledge or scientific truth were limited and muted, unlike the tabloid headline writers and TV news anchors who confidently claimed: 'The God particle has been found!' When asked if he had always been confident that the particle would be found, Higgs replied:

> The existence of this particle is so crucial to understanding how the rest of the theory works as well as it does, in terms of previous experimental verifications of the structure [of particles in the universe], it was very hard for me to understand how it couldn't be there. If it was not there, if it was proved to be non-existent then I would say I no longer understand a whole area of theoretical and particle physics that I did understand.[9]

Higgs's hypothesis has therefore been declared to be scientifically 'true'. However, even this exemplary demonstration of scientific advancement was not without a subjective element: namely, Higgs's own creative input. Whether it is called an insight, a leap of faith, or a logical imagining, his theory was created in an intellectual environment where the boson field did not seem obvious to everyone. The opposite was the case. None other than Stephen Hawking got it wrong, going so far as to bet $100 with a colleague at Michigan University that the particle would not be found. A crucial characteristic of science, whether in the field of physics or climate change, is that some people are proved right by experimentation and observation, while others are proved wrong: personal reputations provide no immunity. Hawking graciously conceded defeat, paid his $100 and congratulated Higgs on his achievement. In such a manner has science always taken faltering steps into the unknown and opened up physical and conceptual horizons that could only have been dreamt of a century ago: space travel, the internet, solar power, text messaging. Furthermore, these achievements have not been made by common consent amongst scientists – even within the same field, as Hawking demonstrated. They have been made when one scientist has shown a hypothesis to be confirmed in his findings, and another scientist has verified the results.

Having introduced the relationship, or at least an idealised relationship, between science and truth, there remains one related matter that will take on greater significance when we turn to examine climate science: the means by which scientists come to be seen, and to see themselves, as ethical. In the pursuit of normal science, scientists emerge as ethical in a relatively straightforward manner. They are required to conform to the established codes and practices of their specialist fields as they combine hypothesis, observation, experimentation and the replication of findings in pursuit of new knowledge, all underpinned by the principles of universalism, communalism, disinterestedness, and organised scepticism expounded by Merton and others. The scientists who observe these standards can be said to act ethically in the pursuit of scientific truth or knowledge, while any scientist who fails to conform to any part of these professional norms is acting unethically. The outcome of scientific enquiry is morally neutral, with a negative result in cancer research being as valid as a positive result. When it comes to the science of global warming and climate change it would be reasonable to expect that the same standards of truth-seeking and ethical conduct will be applied at an individual level. However, the extent to which this is not the case can be surprising, even shocking.

Scientific advancement and truth

Andrew Dessler and Edward Parson, in setting out a case for global warming and climate change, explain how the process of scientific enquiry works. After describing the connection between hypotheses, observation and searching for evidence they go on to make a statement that flies in the face of much scientific discovery: 'A hypothesis that contradicts well-settled knowledge is regarded – reasonably – as almost sure to be wrong, and so is unlikely to attract any interest.'[10] In support of their argument, presumably intended to ward off challenges to what they take to be settled accounts of global warming, they add that 'a new proposal that the Earth is fixed in space and heavenly bodies revolve round it, or that microbial infections do not cause disease, would not attract scientific interest.'[11]

On the face of it these latter comments are uncontroversial and obvious but they skim over some uncomfortable terrain that undermines their generalisations. It is inconceivable that a scientist today would place a static Earth at the centre of the universe on the basis that the claim has been thoroughly tested and rejected over hundreds of years. When, in 1616, Galileo defended his proof that the Earth revolved round the Sun and not vice versa, he was not only contributing to our current understanding of the way that our planet moves, but also undermining established intellectual, scientific and religious 'truth'. Galileo's example is frequently used by opponents of the global warming thesis to show how one person can undermine the entire establishment. They are taking the wrong lesson from Galileo. As far as I am aware nobody has 'proved' that that there has been no human contribution to climate change or that the world did not warm in the final decades of the twentieth century, though some hold that view. Much more interesting is that despite the scientific evidence he offered, in 1633 Galileo was put on trial for heresy for challenging the established, settled truth at that time about the orbit of our planet: for using the nascent modern scientific method to undermine what was seen to be the greater or more important truth about God's created order – truth that was accepted and enforced by institutions including the Church, the universities and instruments of state. Pope Urban had Galileo imprisoned, forced him to recant and had his book *Dialogue Concerning the Two Chief World Systems* banned.[12] Since the printing press was invented, book banning has been a regular feature of disagreements over what constitutes the truth, scientific or otherwise. Note, however, that only those individuals and institutions who wielded considerable political power could enact and maintain a book ban for any substantial length of time.

What Galileo did was not only undermine an apparently absolute truth that had dominated Western thinking for 1500 years. Rather he challenged a 'regime of truth'[13] that relied upon the institutions and teachings of the Church and its authority to declare doctrinal purity or heresy in relation to scientific advancement: all reinforced by political authority manifested in the Holy Roman Empire and beyond. Furthermore, the Catholic Church's Holy Inquisition was used as a tool to suppress any truth claim that sought to undermine the Church doctrine it protected, further restricting the possibility that dissenting opinion could ever hope to be heard, never mind enacted – no matter how scientifically verifiable it was.

If Dessler and Parson's view about hypotheses which contradicts established knowledge being almost certainly wrong is revisited in earlier historical contexts, their claim looks faintly ridiculous, their certainty echoing rather than contradicting earlier 'certainties' that the world is flat and that the Sun revolves around the Earth. As recently as the late twentieth century it was well-established medical orthodoxy that stomach ulcers were caused by stress, diet, stomach acids, or a combination thereof. So when Robin Warren first proposed in 1979 that they might instead be caused by bacteria he was ignored by the scientific mainstream. Normal science had spoken and was no longer listening to novel suggestions. However, in 1982 – in conjunction with colleague Barry Marshall – the bacterium Helicobacter Pylori was identified as the source of almost all duodenal (intestinal) ulcers and gastric (stomach) ulcers.[14] By conforming to the norms and codes of good scientific practice Warren and Marshall were, quite literally, able to stand against a global medical-scientific 'truth' and turn it on its head. As their findings were replicated (again – best scientific practice) their idea rapidly became the new orthodoxy, reinforced by the award of the 2005 Nobel Prize for Medicine. Reason alone, and presumably Dessler and Parson on the reading I have provided, would have maintained the established view on the basis that challenges to it would almost certainly be wrong. However, some leap of intuition, or speculation away from the orthodox view opened up new possibilities.

Eventually, like Warren and Marshall's later work on stomach ulcers, Galileo's scientific claims regarding the relative movement of the Sun and the Earth could no longer be suppressed. Just as he had built on Copernican ideas, so others tested his hypotheses and gradually the old conceptual order gave way to the new: the old regime of truth could only be sustained for so long under pressure from new, demonstrable theory. Thomas Kuhn referred to this process as a scientific revolution,[15] which is subtly different from scientific progress. According to Kuhn, these revolutions have happened throughout history and continue to occur, creating separate, distinct paradigms of thought and scientific practice that are incommensurate with one another.[16] Kuhn's paradigms – which have multiple meanings even in his own use of them – have, however, frequently been over-simplified or misinterpreted, with critics often failing to acknowledge that they include sociological as well as scientific dimensions. He clarifies his definition of 'paradigm' thus:

> On the one hand, it stands for the entire constellation of beliefs, values, techniques, and so on shared by members of a given community. On the other, it denotes one sort of element in that constellation, the concrete puzzle-solutions which, employed as models or examples, can replace explicit rules as a basis for the solution of the remaining puzzles of normal science.[17]

The breadth of Kuhn's concept of a paradigm set out here becomes significant in considering scientific claim and counter-claim regarding global warming and climate change because what can be construed as 'normal science' is so vociferously disputed by climate change advocates and opponents alike: what counts as evidence, on what basis, and under whose authority. The sociological dimension – necessarily subjective since it refers to people, relationships and communities – plays a crucial, frequently unacknowledged role that becomes more apparent as climate debate is analysed. The anthropogenic global warming (AGW) paradigm first emerged in the 1960s and grew to a position of global political and scientific dominance by the twenty-first century. Not just the claim of the existence of AGW and the existence of science-based evidence to support the claim, but also global organisations, professional societies, interest groups, activists, political ideologues and others who make the claim possible and who together constitute and sustain the regime of climate truth on which it rests.

Climate truth and post-normal science

Given the extent to which science, or normal science, has changed the world for the better in recent centuries in terms of food production, transport, advances in communication, medicine, and countless other ways (though I will concede that spin-offs like nuclear weapons, pollution, greed and poverty are more ethically problematic), the casual observer could be forgiven for thinking that there is no need to change a successful formula. Yet away from the scientific and media mainstream that is exactly what has been increasingly happening in recent years in the field of climate science. The basis of this shift is captured in the words of pioneering global warming advocate Professor Stephen Schneider, founder and editor of *Climatic Change* and IPCC lead author. In 1989 he wrote about the changing nature of the relationship between scientific method, truth, ethics and climate change, all with the aim of making the world 'a better place':

> On the one hand, as scientists we are ethically bound to the scientific method, in effect promising to tell the truth, the whole truth, and nothing but – which means that we must include all doubts, the caveats, the ifs, ands and buts. On the other hand, we are not just scientists but human beings as well. And like most people we'd like to see the world a better place, which in this context translates into our working to reduce the risk of potentially disastrous climate change. To do that we need to get some broad based support, to capture the

public's imagination. That, of course, means getting loads of media coverage. So we have to offer up scary scenarios, make simplified, dramatic statements, and make little mention of any doubts we might have. This 'double ethical bind' we frequently find ourselves in cannot be solved by any formula. Each of us has to decide what the right balance is between being effective and being honest. I hope that means being both.[18]

This quote has almost taken on a life of its own since Schneider first made it in 1989. According to prominent climate change advocate Michael Mann, contrarians and deniers – individuals who to different degrees reject the AGW thesis advanced by Mann and many others[19] – like to use the quote selectively, focusing on the line, 'scary scenarios, make simplified, dramatic statements, and make little mention of any doubts we might have,' while avoiding the broader context of Schneider's words.[20] It is obvious why opponents who reject some or all of the climate change argument have seized upon these words, which appear to suggest an incitement to deception, exaggeration and dishonesty. Two recent prominent critics – 'contrarians' in Mann's terminology – James Delingpole[21] and Robert Carter,[22] both refer to this quote and use it in full, as Schneider intended, in the process of criticising what they see as the corruption of science within the climate change community. Probably because it appears to be more, not less, damning when the scientific context of the 'scary scenario' comment is included.

Schneider's words clearly violate aspects of the codes and practices of 'normal' science set out previously: most notably the idea of disinterestedness. However, it is by analysing his words in relation to ethics and truth that their impact becomes even more significant. Practitioners of 'normal' science act ethically as long as they conform to the established scientific codes of universalism, communalism, disinterestedness and organised scepticism.[23] This approach was initially confirmed by Schneider who said that all scientists, 'are ethically bound to the scientific method, in effect promising to tell the truth, the whole truth, and nothing but – which means that we must include all doubts, the caveats, the ifs, ands and buts'.[24] This scientific approach is set alongside a desire to make the world a better place – a value judgement that falls outside the constraints of the scientific method, thereby creating what Schneider calls a 'double ethical bind'. In spite of his claim, there is no double bind here – impressively grand sounding though it is – only an ethical choice. For him, climate scientists who want to conduct themselves ethically are no longer confined to the narrow constraints of 'normal' science: the spectrum of 'ethical' conduct has been extended to include striking a balance between effectiveness and honesty.[25] Schneider's claim to a 'double ethical bind' is misleading, though he does capture the tension experienced by scientists attempting to act ethically in the face of a moral contradiction. What Schneider presents is an ethical choice he is unwilling to make: he wants to pursue what he sees as the ethical goal of making the world a better place, while still being able to claim the moral authority and public cred- ibility associated with normal science, even as he argues that scientific 'norms' would have to be violated in pursuit of higher ideals.

In this highly significant early skirmish in the climate truth wars, Schneider was attempting nothing less than a redefinition of science and the truth that it produces. In his *schema*, the world would be made better by 'working to reduce the risk of potentially disastrous climate change', political actions motivated by particular *beliefs* – that go beyond proven, normal scientific truth – about the world's climate. The crucial point is that the pursuit of Schneider's higher cause of making the world a better place, in his view, justifies the suppression of an important element of scientific truth: namely, the doubts and caveats. The real ethical dilemma for Schneider, and for those who adopt his approach to science and climate activism, is not whether scientific truth claims – traditionally understood – will be violated (he concedes they will) but the *degree to which they will be violated*. The double bind has gone. The politically-informed regime of climate truth takes priority over narrower claims to normal scientific truth and its accompanying qualifications and cautionary tones.

The difficulty for any non-specialist seeking to understand climate change is that when the precision, caveats, constraints, and, most crucially, disin- terestedness of normal science are ignored or diluted, what is left can no longer in any real sense be credibly maintained as purely scientific truth. Schneider's choice – and the choice of others who take his approach – in forming himself as an ethical individual through his conduct and in his beliefs, would appear to be a clear-cut decision between conforming to the codes and norms of the scientific method *or* by creatively extending his understanding of science to incorporate ideological and political ideas and motivations into his moral framework. He prefers not to make that choice, opting instead for an ethical sleight of hand that somehow attempts to provide the basis for a regime of climate change truth in the scientific method, despite the moral contradictions involved: thereby locating himself somewhere 'between being effective and being honest.'[26]

In 1993, Silvio Funtowicz and Jerry Ravetz provided a philosophical framework that would allow Schneider and others to extend the basis of climate study from 'normal science' to include a definite ideological and political dimension.[27] They challenged the dominance of 'normal science' in environmental studies, rejecting the idea that a one-size-fits-all approach to science was still relevant in what they would call an era of 'post-normal' science.[28] The impact that their concept of post-normal science would have on the relationship between science, philosophy and truth was profound. Funtowicz and Ravetz argued that in post-normal science, facts and values should be brought together in problem solving.[29] 'Truth', in the sense of being scientifically demonstrated, would be replaced by 'quality', a subjective value judgement, as a means of assessing the importance of climate science findings.

These ideas, in turn, have been developed further by others. Gert Geominne, similarly challenges the relevance and usefulness of normal science,[30] preferring to distinguish between two aspects of scientific truth: logical truth and topical truth.[31] Logical truth refers here to what is right and scientifically provable, while topical truth refers to what is relevant and of interest. Since topical truth relates to what is of *interest*, say the impact of global warming on the environment, it immediately violates a central element of normal science – disinterestedness. Topical truth is in effect a subjective, qualitative judgement that is claimed as truth. For Geominne, normal science has little topical truth when it comes to environmental and sustainability matters and is therefore deemed inadequate to deal with them and their associated political dimensions.[32] Considered from an opposing perspective, Geominne's 'topical truth' approach politicises climate change and sustainability to the extent that it cannot legitimately be claimed as some extension of science: instead it *contradicts* the codes of normal science. Those who adopt post-normal science as the theoretical basis of environmental science signal the end of normal science in the domain of global warming and climate change, ushering in a regime of truth – specifically a regime of climate truth – wherein scientific truth claims are not shaped by the previously accepted norms but are instead framed by political and ideological pursuits, beliefs or goals.

Dessler and Parson set out a case for checks and balances in the search for scientific knowledge – which is not for them a 'proven truth' – in the peer-review process, whereby the validity of claims and findings is scrutinised and tested.[33] Their rejection of the idea of 'proven truth' contrasts with the high degrees of certainty claimed by climate alarmists and contrarians alike. Peer-review underpins the pursuit of objectivity in science, up to the point where they refer to theoretical claims or observations and measurements being 'judged by the *relevant* scientific community.'[34] The question of *who* is allowed to pass scientific judgement on others is to some extent a subjective, not simply an objective assessment, and can be open to abuse. Of course, nobody is ever going to own up to unscrupulous activity – at least not to a conscious, deliberate intention to mislead. In theory, science is science is science, so it does not matter who conducts peer reviews. Or does it? It takes a special kind of person to negatively critique a view or finding that they personally hold dear and on which their career, salary funding and credibility depend, regardless of whether that funding comes from the oil industry, governmental research grants, or climate change advocacy groups. Nobel prize-winning scientist Professor Randy Schekman, a cell biologist, has strongly criticised academic journals, especially what he calls prestigious, 'luxury-journals', for widespread, questionable editorial policies that undermine good science.[35] He is especially critical of the practice of publishing papers on 'sexy subjects' or which make bold claims – which may not be supported by the evidence on offer – while marginalising studies that replicate the findings of existing research: a crucial scientific function.[36]

The peer-review process, valuable though it is in every field of intellectual endeavour, is not quite the guarantor of the quality of specific knowledge claims and academic advancement that many scientists would assert. At its worst, the peer-review process constrains what can be considered to be valid knowledge in the first place and by whom: it plays a crucial role in creating and maintaining regimes of truth in specific and specialised subject areas. The more specialised the area, the greater the possibility that the peer-review process might simply reinforce the expectations of the group it serves. Practically speaking, such an outcome is almost inevitable since a narrow field of inquiry will necessarily draw upon a smaller pool of potential peer reviewers. As the discussion moves on to explore aspects of the way that climate science works to make and reinforce truth claims in relation to normal and post-normal science, we will return to this question of peer-review and its objective and subjective dimensions.

Constructing a regime of climate truth

In the climate truth wars, the behaviour of some climate scientists in recent years demonstrates a distinct commitment to sustaining a regime of truth through the exercise of individual and collective political power and influence, often with the best of intentions, that goes beyond a mere desire to push back the boundaries of scientific knowledge. Most famously, evidence of such behaviour emerged during the Climate gate controversy, when a collection of climate research-related emails and papers were hacked from the Climatic Research Unit (CRU) at the

University of East Anglia in the UK.[37] I have chosen the following examples to illustrate subjective influence on the production of truth in the process of climate science, not to try and prove whether any one particular outcome is right or wrong.

On 31 January 2003 Willie Soon and Sallie Baliunas, from the Harvard-Smithsonian Centre for Astrophysics, had a paper published in the peer-reviewed journal *Climate Research,* which supported the existence of a Medieval Warm Period – contradicting and challenging the views of a number of leading global warming advocates.[38] What happened next highlights the extent to which climate debate has departed from established norms of scientific practice – setting out the evidence and demonstrating a hypothesis, before having those claims interrogated by one's peers – towards an orchestrated attempt to create a regime of climate truth, all reinforced by the actions of institutions and a collective of like-minded scientists. Consider excerpts from an email sent by Michael Mann to Phil Jones, two of the world's most high profile climate scientists:

> The Soon & Baliunas paper couldn't have cleared a 'legitimate' peer-review process anywhere. That leaves only one possibility – that the peer-review process at Climate Research has been hijacked by a few skeptics on the editorial board ... It is pretty clear that the[s]e skeptics here have staged a bit of a coup, even in the presence of a number of reasonable folks on the editorial board (Whetton, Goodess, ...).
>
> So what do we do about this? I think we have to stop considering 'Climate Research' as a legitimate peer-reviewed journal. Perhaps we should encourage our colleagues in the climate research community to no longer submit to, or cite papers in, this journal. We would also need to consider what we tell or request of our more reasonable colleagues who currently sit on the editorial board ... What do others think? Mike[39]

There is no mention in this communiqué of the content of the published paper: the whole response was a concerted attempt to undermine the credibility achieved by Soon's and Baliunas's peer-reviewed journal article by discrediting the publication and its officers. This front in the climate truth wars owes little to contesting scientific truth claims and a lot to institutional politics: discredit the journal – discredit those who publish in it. In a military confrontation this would be referred to as shaping the battlefield, ideally defeating an enemy before a shot is fired once they realise that they have been and will continue to be outmanoeuvred and outgunned. So what was the perceived problem and how was this problem to be solved?

Mann starts off by questioning the legitimacy of the *Climate Research* journal peer-review process itself, and by implication the scientists who conducted the reviews of the offending article and made recommendations for publication. He offers no supporting evidence as to the scientific qualifications or experience, or lack of qualifications and experience, on the part of those who reviewed the Soon and Baliunas paper. Instead he attacks their character, judgement and professional perspective: all summed up in the derogatory term – at least derogatory in the sense he used it – 'skeptic'. By implication, sceptics are unreasonable and wrong, unlike those who agree with Mann's perspective on the world, AGW and climate change. (Set aside for now the centrality of scepticism in normal science – scepticism in climate science is only allowed if it is takes place within the regime of climate truth and does not seek to undermine it.)

Mann's highly personalised approach is intentionally divisive: anyone who did not or does not support Mann's stance is against him and his like-minded colleagues. The simple binary choice he advocates has some advantages. It forces people to commit to one position or another and avoids dealing with the subtle nuances of the middle ground in the climate debate – those who want more information before committing to a position. It echoes a more strident call to arms made by President George W. Bush several days after 9/11 when he asked the world to make a stark choice: 'Either you are with us or you are with the terrorists.'[40] This is hardly the intellectual or methodological legacy bequeathed by Galileo, Isaac Newton or Albert Einstein. And if anyone thinks I am comparing Mann to an unnecessarily extreme example of divisive political discourse, look no further than the title of his own book: *The Hockey Stick and the Climate Wars*. Global warming and climate change – assuming for now that Mann is entirely, unquestionably correct in his view – are not simply challenges to be addressed in terms of scientific findings: those who disagree with him have to be defeated in some kind of 'war'. He talks of climate wars: I see truth wars. His reference to a 'coup' in the email above demonstrates the militaristic language and sentiments that have characterised Mann's approach – and the approach of countless other alarmists and sceptics – for several years.

The problem as Mann presents it has a straightforward solution. Ignoring the possibility of rebutting contradictory and inconvenient evidence scientifically, Mann's proposal is to get like-minded and influential scientists together for a full-frontal assault on the existence of the journal – or at least on its reputation – that was foolish enough to publish a breadth of scientific opinion, something that is often, perhaps naïvely, taken as an academic journal's *raison d'être*. Articles would no longer need be submitted and citations should be avoided, thereby undermining the credibility and academic life-blood of the offending journal. Fellow scientists would be categorised as either 'reasonable' or 'unreasonable' according to their sympathy towards Mann's position ('with us or with the terrorists'). Any objective scientific refutation of the claims by Soon and Baliunas was relegated behind the subjective political machinations and attempted coercion and control of a journal's editorial policy that characterised Mann's attempt to collude with like-minded colleagues to reinforce a particular regime of climate change truth. Further, the successful maintenance of such a regime of truth would remove any need to provide a scientific rebuttal anyway. Paradoxically, such a course of action demonstrates a lack of confidence in good scientific practice whereby disreputable articles would be discredited in due course in the normal scientific scheme of things.

Mann would later dismiss any counter-claims by the opponents of dangerous AGW on the basis that they do not publish their research findings in peer-reviewed journals and that they operate outside established scientific processes for generating knowledge.[41] Such a claim certainly appears to contain a degree of truth but what Mann is less forthcoming about is the reason for the lack of participation: the actions of those within a self-referential climate 'community' deliberately restricting the opportunities for contesting voices to be heard. Such behaviour should be more rightly classified as politics and not science. Undermining his own argument Mann also, ironically, notes that the peer-review process is no guarantor of research quality anyway.[42] His approach seems to be: just in case there is a chance that your opponent's hypothesis may well be proved correct, true or at least statistically feasible, or may have undue influence on the those who are easily swayed, use other methods to ensure that it is marginalised and left marooned outside a politically and socially constructed regime of climate truth. Foucault's maxim that politics is a continuation of war by other means finds credence in the climate truth wars. Mann and fellow scientists who are warning of the dangers of unchecked AGW may well be proved completely correct in their analysis, with their position continually confirmed and reinforced in the coming years. However, trying to enforce methodological boundaries within a particular regime of climate truth may well result in the unintended consequence of unwittingly silencing an awkward voice somewhere that is trying to make the world aware of some vital unseen connection, or break a conceptual link that has been made in error. Another unintended consequence might be the alienation of potential supporters whose openness to persuasion by climate science argument is replaced by a mistrust of political manoeuvring. The reasoning goes like this: if you have to resort to those tactics then you must not have a good argument in the first place.

Attempts by a number of climate change advocates to pressure the journal *Climate Change* into altering its editorial policies was not an isolated case. Two years later the CRU emails highlight another example where a political fix was being proposed to solve the 'problem' of dissenting scientists having their papers published. At the start of 2005 another scientific journal, *Geophysical Research Letters* (GRL), published an article by Steve McIntyre that challenged some of Mann's earlier work on the Hockey Stick global temperature graph. Before doing so, however, the journal's Editor in Chief, Steve Mackwell, emailed Mann and asked if he would like the opportunity to respond to McIntyre's claims in typical fashion where there is a scientific dispute.[43]

Mann's subsequent email to like-minded colleagues and supporters made no mention of any scientific rebuttal of his interlocutor's claims. Instead it highlighted his concerns about his opponents even having a voice:

> Dear All,
>
> Just a heads up. Apparently, the contrarians now have an 'in' with GRL. This guy Saiers has a prior connection w[ith] the University of Virginia Dept. of Environmental Sciences that causes me some unease. I think we now know how the various Douglass et al papers w[ith] Michaels and Singer, the Soon et al paper, and now this one have gotten published in GRL, Mike[44]

Mann has moved on from the use of 'skeptics' to describe individuals who challenge his views to 'contrarians'. There is good reason for the change. While his use of 'skeptic' in the 2003 email exchange quoted above was intended in a derogatory sense – as is the use of 'contrarian' here – it is also problematic. Scepticism and sceptics have always played a crucial role in challenging the *status quo* within science, sometimes leading to unforeseen and highly significant new ideas and under-standing. One of the aims of scientists who were and are convinced that dangerous human-induced global warming exists and must be ameliorated is to cut off further debate lest doubt starts to set in – especially if global mean temperature stops rising for any length of time as it has since 1997 – and support for political intervention begins to wane. Harsh critics of those climate scientists, like James Delingpole (journalist and blogger), frequently resort to rather uncharitable language and are summarily dismissed or ignored by his targets:

> The people who tell you that AGW is a near-certainty are a bunch of liars, cheats and frauds. Your taxes will be raised, your liberties curtailed and your money squandered to deal with a 'crisis' so exceedingly unlikely and so poorly supported by real world data or objective science that it might just as well not exist.[45]

Ignoring critics and criticism is not the same thing as proving them wrong. Other critics are harder to ignore, like Professor Robert Carter, a paleoclimatologist with 40 years' experience researching climate and the natural environment, who readily acknowledges a human contribution to global warming. However having studied climate in a geological context going back millions of years he is of the view that recent warming falls within the natural variability of the Earth's climate. Furthermore, he considers that the man-made warming contribution to recent climate change and a fixation with inexorable global temperature increases mean that inadequate consideration has been given to the possibility that natural variability could even result in a temperature fall this century: for him a much more damaging proposition than warming. He states that

> the human signal most probably lies buried in the variability and noise of the natural climate system. This is to such a degree that as a statement of fact we cannot even be certain whether the net human signal is one of global warming or global cooling. Though it is true that many scientists anticipate on theoretical grounds that net warming is the more likely, no direct evidence exists that any such warming would *ipso facto* be dangerous.[46]

Carter's approach appears to be both reasoned and reasonable: he acknowledges that humans are probably contributing to global warming, while urging caution on the level of supporting evidence, the likely outcome for life on Earth and the actions that should be taken. His very reasonableness is, however, the problem. To use a legal phrase, he challenges the notion that the existence and threat of dangerous AGW is beyond reasonable doubt. Yet doubts and caveats, the 'ifs, ands and buts' – if we recall the words of Stephen Schneider earlier in the chapter[47] – are what must be banished from global warming discourse if politicians and the public are to be convinced by the regime of climate change truth and prompted into action.

Having explored how the basis of climate science has shifted from 'normal' science to 'post-normal' science, it is apparent that the very scientific basis of climate change and global warming remains disputed, with highly subjective influences shaping research and discussion at every level. Chapter 2 will look at subjective aspects of the politics of climate debate where scientists' claims meet the world of climate activism. It will become clear that there is risk involved in a shift away from the narrow confines of normal science towards ideologically-motivated and politically-shaped post-normal science as the basis of a regime of climate truth, especially when it comes to proposing action to remedy climate change. One threat to the claims of climate change advocates comes from those who embrace the enduring, and more limited, principles of normal science and the public credibility it still holds. The nature of that threat to the regime of climate truth that has emerged since the 1980s will be explored in the pages to come.

2
Politics and Climate Truth

The ferocity of the climate truth wars reached new levels when the ideological battles ventured beyond the scientific domain to the realm of political activism, with each side aiming to shape individual behaviour and gain support for its particular view of how the natural world should look both in the present and the future. A regime of climate truth has increasingly shaped local, national and international politics over the past three decades, with ideological assumptions and political goals all underpinned by post-normal scientific practices. In the process, climate change and its advocates have come to dominate the political landscape, marginalising – but not silencing – dissenting voices in the process.

This chapter explores what happens when climate science is conflated with political and ideological beliefs and goals in the name of truth. Discussion will concentrate on events in the UK where climate change truth claims have extensively influenced public perception and government policy, which plays what Energy Secretary Ed Davey describes as 'a leading role' in reforming the European climate change and energy framework.[1] To be clear from the outset, I reject the claims of climate change deniers and sceptics who, in turn, reject global warming out of hand (global mean temperature increases in the late twentieth century have been independently measured by three separate and credible internationally regarded organisations). Further, I also reject as compromised and illegitimate any anti-global warming or anti-climate change research funded by carbon generating industries. Much more interesting is an exploration of the ways in which climate change truth is constituted and deployed in political discourse through the use of mass media, education policy and other government institutions and, further, how the way climate truth is created leaves it vulnerable to attack by the very opponents it seeks to undermine.

The first section will consider how the dominant understanding of climate truth has changed in recent decades, with global warming and climate change ceasing to be independent areas of disinterested scientific research and instead being subsumed into Climate Change, an ideology whose emphasis has similarly shifted from 'proof' to 'belief'. This approach is rejected by some in the field who seek to exclusively uphold the codes of normal science, and who are less comfortable about surrendering those codes in pursuit of the ideological claims associated with post-normal climate science. The second section will examine how a regime of climate change truth has extended its reach, a crucial element of which is gaining control of significant media outlets, with the highest profile and most complete example of a national broadcaster being recruited to the climate cause in recent years occurring in the UK with the BBC. Subsequently, an examination of political activism in support of a regime of climate truth will demonstrate not only its effectiveness when it comes to shaping attitudes and behaviour, but also its potential vulnerability. Then finally, a high profile and controversial example will illustrate, first, the extent to which a post-normal science-inspired regime of climate truth has come to dominate the climate truth wars, thereby dominating the political landscape; and second, that there is a major vulnerability at the heart of Climate Change.

Climate change as political ideology

Mike Hulme, high profile climate professor and public commentator, places climate change at the heart of a vast global project with political aspirations that range from shaping the way individuals live their lives to the redistribution of wealth:

> The idea of climate change should be used to rethink and negotiate our wider goals about how and why we live on this planet. We need to harness climate change to give new expression to some of the irreducible and intrinsic human values that are too easily crowded out – our desires for personal growth and self-determination, for creative experimentation, for relationship and for community... The function of climate change I suggest, then, is not as a lower-case environmental phenomenon to be solved... It really is not about stopping climate chaos. Instead we need to see how we can use the idea of climate change – the matrix of ecological functions, power relationships, cultural discourses and material flows that climate change reveals – to rethink how we take forward our political, economic and personal projects over the decades to come.[2]

Hulme articulates what many climate change advocates suggest but do not state so openly or explicitly and which opponents have accused them of for years. Climate change is no longer just about – if it ever was – overcoming the physical challenges of global warming and associated changes in the Earth's climate. For Hulme and others who share his worldview, climate change needs to be capitalised – Climate Change – to acknowledge that it has become a proper noun – a 'thing', an ideology, a political movement. Climate Change has an existence of its own, sustained by a regime of climate truth that is constructed and maintained within countless relations of power. Furthermore, the focus of Hulme's statement is almost entirely *subjective* (based on opinions, perceptions, values, power relationships and personal aspirations) rather than *objective* (concerned solely with measurement, evidence and proof). 'Belief' has become the defining characteristic of the Climate Change proponent. Consequently, from here on Climate Change will be used in its capitalised form where I refer to a post-normal science-based, ideologically informed political movement, rather than mere physical changes to the climate or global mean temperature. In addition, I will avoid the established oppositions of Alarmist, Sceptic, Contrarian and Denier when describing individual relationships with Climate Change and with each other. Instead, I will use terms like adherent, supporter, believer, or doubter, opponent and agnostic (the latter thinks that not enough is yet known about all the variables of climate science to form a valid opinion).

Climate Change inverts the relationship between science and politics. True Climate Change advocates are no longer simply examining a physical scientific problem then asking how, politically and economically, it can be solved: understanding climate change – its extent and its causes – is no longer the primary objective of their study. The object of their endeavour is now shaping human lives in what they see as problematic political-economic ideology, with Climate Change – 'the matrix of ecological functions, power relationships, cultural discourses and material flows that climate change reveals'[3] – being used as the means to achieve it. Ask not what you can do for climate change; instead, ask what Climate Change can do for you (with apologies to JFK). Reinforcing my reading of Hulme here is the declaration that he is 'fascinated by the malleability of the idea of climate change as it came to be appropriated in support of so many causes.'[4] The causes are the priority, with Climate Change used to support them. Put more simply, Climate Change 'truth' no longer refers just to scientifically-observed changes in the climate but now also acts as a new front in the ongoing ideological struggle between the political Left and Right. The Left appropriates climate change as a means to advance a global redistribution of wealth – which is also seen as a legacy of colonialism and imperialism for which restitution should be paid. On the Right, some deny the existence of climate change or global warming not on scientific grounds but because of the political associations. Others accept there has been a human contribution to global warming and climate change but consider any deleterious effect to be insignificant compared to the improvements that technology, science, medicine, communication and countless other developments have brought to human existence. While another broad group on the Right seeks to gain support by ameliorating the worst climate and environment-related excesses of human behaviour while still advancing capitalism.

Climate Change is therefore to be used for political ends, not solved by political means. The beauty of such an approach is that there is no endgame. As Robert Carter puts it, '*Change is simply what climate does*'.[5] Therefore, the need for political action will never go away, and neither will the need to shape people's behaviour with respect to the regime of Climate Change truth: it is a tool for the control of populations that will never tire, regardless of whether the global mean temperature is rising, falling, or, as it has been between 1997 and 2014, standing still. That is why outgoing British Government Chief Scientific Advisor, Sir John Beddington, could say in a television interview in 2013, 'We have moved from the idea of global warming to the idea of climate change.'[6] One *idea* has replaced another. It is not a case of one physical challenge replacing another; there has been a political and ideological shift. The relative unimportance of the 'truth' about what global temperatures are actually doing was illustrated when he went on to add, 'though global temperatures are still rising.'[7] In a similar vein, at the launch of a new Intergovernmental Panel on Climate Change Report on 31 March 2014, Michel Jarraud, World Meteorological Organization Secretary General, said that since thirteen of the previous fourteen years had been the warmest on record: 'I really refuse to accept that we can talk about a pause.'[8]

There are two ways of interpreting Beddington's statement that global temperatures are continuing to rise and Jarraud's rejection of a pause in global warming, each of which might be referred to as 'truth'. If they are comparing the average, measured global temperature over the previous ten years (say, 2003 to 2013) to the average global temperature of the twentieth century then it is demonstrably 'true' that the recent ten-year average is higher. In contrast, if they are implying – and the delivery of their words suggests this – that the global mean temperature is continuing an inexorable and unimpeded rise, then this claim – in the sense of scientifically verifiable truth – is not credible and offers ammunition to their critics. Problematically for Climate Change advocates, for two such key figures – among many – to ignore or skirt over a 16-year standstill in mean global temperature rises provides an opportunity for opponents to claim that there is something to hide. The other way that Beddington's and Jarraud's comments can stand unquestioningly as 'truth' is if they are speaking from an ideological perspective in which continuing, unimpeded, rising global temperature is an article of faith or ideological belief rather than simply a matter of scientifically verifiable fact. And if this is the case, Climate Change ideology should be publicly acknowledged rather than disguised by a description of the current behaviour of global mean temperature that is not reflected in the temperature measurements used by the IPCC.

Elsewhere, Mann, like Hulme and Schneider before him, acknowledges a political dimension to Climate Change that goes beyond the scientific aspect of the climate truth wars. He seeks to use the moral credibility and authority associated with normal science in the political arena, without explicitly stating that he has eschewed the strict practices of normal science in favour of the political dimension of post-normal science. He writes: 'Scientific truth alone is not enough to carry the day in the court of public opinion. The effectiveness of one's messaging and the resources available to support and amplify it play a far greater, perhaps even dominant role.'[9] His words highlight a tension between rigour and relevance as he helps to constitute a regime of climate change truth by advocating means that go beyond merely expounding narrow scientific truth claims. In addition, his use of the language and imagery of battle displays the extent to which he will wage this war of ideas: 'When scientists are willing to fight for their cause – in this case, communicating the potential climate change threat – there are many good men and women who will not stand by and do nothing.'[10]

Mann's personal ethic and his self-identity as a human and as a scientist is rooted in Climate Change, a cause which he advances by communicating the threat to the planet associated with man-made climate change. In 2004 he articulated his personal and professional commitment to the climate cause in an email – one of the hacked 'Climategate' emails – to Phil Jones at the Climatic Research Unit:[11] 'By the way, when is Tom C going to formally publish his roughly 1500 year reconstruction??? It would help *the cause* to be able to refer to that reconstruction as confirming Mann and Jones, etc.'[12] Although Mann undertakes scientific research, like Hulme, it is the climate *cause* that emerges here as his ethical priority, not merely the objective replication or confirmation of his own research findings.

Not all climate scientists, however, including a number who advance the dangerous anthropogenic global warming thesis, are happy about the way that science has been and is being used for political ends. A sense of concern, anxiety even, can be detected in some of the Climategate emails, where private reservations within the scientific community were laid bare. Peter Thorne, climate scientist and lead author on the IPCC 5th Assessment, Chapter 2: *Atmosphere and Surface Observations*,[13] expressed his disquiet in 2005. The context of his comments was the preparation of the draft text for AR4, the *IPCC Fourth Assessment Report: Climate Change 2007*. He was of the opinion that one of the papers to be used in the Report was simply wrong and 'may make us all painted [sic] into a difficult corner.'[14] Furthermore, he added:

> I note that my box on the lapse rates was completely and utterly ignored which may explain to some extent my reaction [in a previous email exchange], but I also think the science is being manipulated to put a political spin on it which for all our sakes might not be too clever in the long run.[15]

Thorne was concerned that his contribution to the assessment paper was being ignored, the impact of which would be a reduction in the scientific accuracy of the final Report. More important than his unease about an erroneous scientific paper was the reason behind the selective use of climate data and findings: the science being manipulated for political reasons.[16] Thorne's email, however, should not be read as some 'smoking gun' that reveals fraud or deception in the climate science community. Instead it exposes the tension experienced by someone caught between a desire to conform to the norms and codes of established, normal scientific practice and the political ends to which tentative or qualified postnormal scientific climate change findings were being put. Furthermore, his recognition that the flawed relationship between science and politics may result in future negative consequences was accurately perceptive for reasons that will be explored in Chapter 3.

The shaping of climate truth

While Thorne is attempting to preserve scientific norms, a different dynamic emerges in some of Phil Jones's Climategate emails. Jones appears willing to ignore or even violate the codes and norms that have sustained scientific enquiry for centuries as he helps construct and maintain a regime of Climate Change truth. Given that a fundamental element of scientific practice is the replication of results,[17] observe Jones's response to a request for raw climate station data from Warwick

Hughes: 'We have 25 or so years invested in the work. Why should I make the [climate] data available to you, when your aim is to try and find something wrong with it? There is IPR [Intellectual Property Rights] to consider.'[18] Historically speaking, within normal science the whole reason for testing results and replicating findings is to try and 'find something wrong with it', and Jones goes out of his way to prevent this from happening. He uses intellectual property rights as a justification for not providing data that Hughes wants to test – an odd defence given the amount of public funding that has been poured into the Climate Research Unit where Jones works. Elsewhere, Jones expands on his unwillingness to provide information to inquisitors who would examine and test it: 'If [opponents of Jones's work on climate change] ever hear there is a Freedom of Information Act now in the UK, I think I'll delete the file rather than send to anyone...We also have a data protection act, which I will hide behind.'[19] Jones is stepping outside the norms of scientific practice by obstructing attempts to test and replicate results, apparently with the intention of supporting his ethical priority of maintaining a regime of Climate Change truth in the face of mounting resistance. The climate truth wars were getting ugly and personal.

James Delingpole is scathing in his personalised criticism of Climate Change scientists and the way they work:

> We see them 'cherry-picking' data that supports their theories and burying data that doesn't. We see them drawing conclusions based on gut-feeling rather than evidence. We see them ganging up to bully editors, journalists and fellow scientists who disagree with them. We see them orchestrating smear campaigns. We see them subverting and debasing the peer-review process. We see them insert bogus graphs and misleading information into official reports which are supposed to represent the 'gold standard' of international scientific knowledge.[20]

Delingpole pulls no punches in venting his considerable ire about climate science and climate scientists. Given that he uses climate scientists' own words in his arguments against them, probably the most serious charge that can be laid against him is that he takes select elements of emails out of context. In that regard he can be accused of the same kind of cherry-picking he was accusing his rivals of. Tellingly however, I am not aware of any legal writ being taken out against him for libel. Consequently, the relevant question is not: Are the private communications between climate scientists consistent with their public statements and publications? Most people would not like their office conversations, arguments, innuendo and disparaging remarks appended to every document or project they worked on. The more interesting question is: *Why* might there be a discrepancy between the uncertainties found in private emails and the bold certainties proclaimed in public climate change discourse? The answer is that the discrepancies are a necessary by-product of the Climate Change cause. Normal science-based truth claims are limited and usually qualified. They are therefore unable to support the more certain, sweeping ideologically-informed truth claims that characterise the political dimension of Climate Change.

Take another example from Phil Jones in 2004, who wrote the following in a discussion about vapour pressure, humidity and temperature: 'Basic problem is that all models are wrong – not got enough middle and low level clouds. Problem will be with us for years.'[21] There is nothing controversial about such a statement. By definition all models must be wrong because they cannot include every potential variable in complex systems that they are often called upon to explain, and few are more complex than the biosphere that is planet Earth. In strictly scientific terms, the only legitimate debate is over the *extent* to which any particular model might be wrong, or at least limited, and hence to factor into future projections appropriate error factors. The difficulty is that if a scientist does not know how inaccurate her model is, she must similarly be unaware of the potential magnitude of the error it will produce. To paraphrase the words made famous by Donald Rumsfeld in justifying the 2003 Iraq War, it is the unknown unknowns that will catch us out. Which brings us back to the climate truth wars and the misplaced certainty portrayed by protagonists on both sides. Acknowledging problems and difficulties in one's arguments – as well as accepting accuracies and valid statements by the climate enemy – is usually seen as some kind of weakness that might in itself undermine the entire Alarmist or Sceptic edifice.

The following dispute over climate variables demonstrates a reluctance to incorporate complicating or unhelpful elements into the regime of Climate Change truth. On 27 December 2006 a discussion was ongoing between scientists Curt Covey, Fred Singer and journalist Christopher Monckton, linking together the issues of polar ice melt, the influence of the sun on Earth's climate, and the lack of global warming between 2001 and 2006:

> Mean global temperature has hardly risen at all in the five years since the IPCC's last report. And the fact of the 20th-century temperature increase tells us nothing of the cause. It is interesting, for instance, that the polar icecaps on Mars are receding, inferentially in response

to increased solar activity. At any rate, it is certain that anthropo-genic planetary warming is not responsible. It is possible, therefore, that most of the warming [on Earth] both before and after 1940 was heliogenic.[22,23]

There appears to be nothing controversial about this statement in the context in which it was made. An anomaly had been spotted – global temperatures that rose in the late twentieth century had been observed to have stopped rising for several years in the early twenty-first century – against the predictions of previous global warming models. Monckton had identified a phenomenon whereby Martian polar icecaps may have melted in response to increased solar activity and raised the question of whether the Sun could be having a similar impact on Earth's polar icecaps. Exploring and hopefully solving or explaining anomalies is what good science does and can provide the creative impetus that leads to new possibilities, thereby advancing existing knowledge. It also chal-lenges accepted climate scientific truth. Eventually the above email trail was copied to Michael Mann. Mann's response to what seems like a legitimate line of scientific questioning captures the hostility of the climate truth wars in a highly personal attack:

> Curt, I can't believe the nonsense you are spouting, and I furthermore cannot imagine why you would be so presumptuous as to entrain me into an exchange with these charlatans... I find it terribly irrespon-sible for you to be sending messages like this to [climate skeptics] Singer and Monckton. You are speaking from ignorance here, and you must further know how your statements are going to be used. You could have sought some feedback from others who would have told you that you are speaking out of your depth on this. By instead simply blurting all of this nonsense out in an email to these sorts [of] charlatans you've done some irreversible damage. [S]hame on you for such irresponsible behavior! Mike Mann.[24]

Whatever reputation science and scientists might have for detailed analysis, careful experimentation, cautious declarations of find-ings, and so on, it is not reflected in Mann's response. His vocabu-lary is derogatory towards his opponents and anyone who engages positively with the climate enemy: 'Nonsense... spouting... presump-tuous... charlatans... ignorance... out of your depth... blurting... irrevers-ible damage... irresponsible behaviour'. Mann is content to ignore a potentially problematic question: What if? What if solar activity plays a significant but yet undefined role in altering global mean temperature? For Mann, to consider that question is not to extend climate knowledge but to do 'irreversible damage'. That damage cannot be to the integrity of science – disproving a hypothesis contributes to the production of scientific truth just as much as confirming a hypothesis. The damage to which Mann refers is to the regime of climate truth and the Climate Change cause that he supports so passionately. Again, the fault line within that regime of truth comes to the fore, with the pursuit of legitimate scientific enquiry being subordinated – by a scientist in this case – to broader politically-motivated ideological considerations.

In *The Hockey Stick and the Climate Wars*, Mann sets out his own formula for shaping public perceptions of Climate Change: scientists must work alongside social scientists, public policy makers, public relations experts, sympathetic financial foundations, suitably interested private sector institutions, and social media and internet-based campaign groups.[25] The primary purpose of this collective of disparate groups, according to Mann, would be to achieve what he sees as more accurate and more objective coverage of climate change matters in the mainstream media.[26] However, his apparent pursuit of objectivity is not made in the normal scientific sense. Mann seeks to change perception rather than understanding, a subtle yet crucial distinction. A public relations company adds nothing to a scientific understanding of climate change, but it *can* help to shape how a particular message on climate will be received by listeners. Mann, therefore, takes a positive view of institutions and individuals, scientific or otherwise, who are allies in the production and promotion of climate truth. For example, he is supportive of the *Washington Post*'s willingness to 'do their part'[27] for the cause, not by adding to the scientific debate but by promoting politically-oriented Climate Change 'truths' with which he agrees.

Referring the discussion so far back to the conception of truth I set out in the Introduction, consider the words of Michel Foucault on truth as socially constructed within relations of power. He wrote:

> Each society has its régime of truth, its 'general politics' of truth: that is, the types of discourse which it accepts and makes function as true; the mechanisms and instances which enable one to distinguish true and false statements, the means by which each is sanctioned; the techniques and procedures accorded value in the acquisition of truth; the status of those who are charged with saying what counts as true ... 'Truth' is centred on the form of scientific discourse and the institutions which *produce* it ... it is produced and transmitted under

the control, dominant if not exclusive, of a few great political and economic apparatuses (university, army, writing, media) [and] is the issue of a whole political debate and social confrontation ('ideological struggles').[28]

Mann's proposals are almost a perfect summary of how to create a regime of truth in a manner hypothesised by Foucault before words like climate change, internet and International Panel on Climate Change became part of the global lexicon. Climate truth is produced using institutions, skills and interests that span a whole range of professional expertise: all focused on producing an ideologically-driven view of the world that will, in turn, shape individual behaviour within power relations as a result of political decision making. Mann understands, even if he does not articulate it in these terms, that the 'truth' about global warming and climate change does not have some scientific essence of its own that can somehow speak up for itself. Climate truth, as he, his allies and his enemies in the climate truth wars all understand, is *made* to count as true in public discourse, using institutional mechanisms and social procedures. Mann has identified those necessary mechanisms: from sympathetic funding sources to social media campaigns to mainstream media coverage and, ultimately, shaping political policy. Once Climate Change is taught in schools and universities as factual and only contested by the mad, bad or sad – but without explicit mention of its ideological dimension – and something to be stopped through political intervention, it becomes very difficult for doubts to be aired and heard: whether those doubts are about the scale of the danger or how that danger should be addressed. It becomes even more difficult when mass media is brought into the Climate Change fold, with its capacity to shape opinion through the broadcasting of particular truth claims. In the early years of television the medium was used for what we now call 'infomercials': often health focused and giving guidelines about vaccinations, general hygiene or – at the height of the Cold War – what to do in the event of a nuclear attack. For those same, behaviour-shaping capabilities, global advertising is a multi-billion pound industry. Less subtly, every despot knows that control of the media is crucial to control of the people and every campaign manager in democratic elections knows the value of effective press management.

Broadcasting climate truth

In the UK, the extent of popular acceptance of the threat of global warming and climate change placed it, alongside Germany, at the forefront of European climate policy formation, and well ahead of the US, mainly because of the way various necessary elements of a regime of climate truth have come together. Perhaps most crucially, the British Broadcasting Corporation (BBC) was successfully brought into the Climate Change fold with little fanfare and great efficiency, thereby allowing one voice to dominate British mainstream media.

In February 2011, retiring veteran news anchor Peter Sissons denounced the way pro-AGW voices had come to dominate the news agenda in the 2000s, saying that the 'BBC became a propaganda machine for climate change zealots'.[29] His experiences and observations capture in some detail the extent to which the BBC was brought into the regime of climate truth to support the Climate Change cause. Recognising the difference between science and activism, Sissons observed: 'Environmental pressure groups could be guaranteed that their press releases, usually beginning with the words "Scientists say ... " would go on air unchallenged'.[30] There are three important aspects to this brief statement. First, it was environmental pressure groups that were doing what their names suggest: putting pressure on the BBC to present only a pro-Climate Change position. Second, the pressure groups would seek credibility for their claims by stating that they rested upon scientific evidence – with no mention of post-normal science and the ideological assumptions it contains. Third, the ideas put forward by the pressure groups would go unchallenged, thereby contradicting the BBC's regulatory obligations to 'do all it can to ensure that controversial subjects are treated with due accuracy and impartiality in all relevant output.'[31] The BBC would become very partial in climate matters.

The BBC's adherence to regulatory guidelines on impartiality began to officially give way to overt support for a regime of climate truth with effect from 26 January 2006. A day-long 'high-level seminar' entitled, 'Climate Change – The Challenge to Broadcasting', was held at the BBC Television Centre under the direction of Roger Harrabin, a BBC environmental reporter, and Dr Joe Smith of the Open University. The BBC would later refer back to this event as its authority for subsequent uncritical prioritisation of pro-Climate Change discourse. A 2007 report on impartiality stated: 'The BBC has held a high-level seminar with some of the best scientific experts, and has come to the view that the weight of evidence no longer justifies equal space being given to the opponents of the consensus.'[32] What was meant by 'consensus' was not explained and the qualifications and expertise of those making up the consensus were similarly omitted from the report. The implication was that since the seminar comprised 'the best scientific experts', unimpeachable scientific credentials and evidence stood behind the BBC's new pro-climate change stance. No objective scientific basis for the BBC's position was publicly offered, its new policy resting entirely upon implied subjective evidence: the credibility of the scientists in question. The Climate Change cause and its regime of climate truth had expanded to include a state broadcaster with global reach.

Over time, however, the credibility of the BBC's policy decision – and the credibility of the scientific gathering that supported it – came under attack. On 14 December 2008, journalist Richard North described his experience of the BBC's climate seminar as 'shocking', pointing out that there were as many BBC representatives as climate specialists, the vast majority of whom were climate activists rather than scientists.[33] Having made the major policy decision to promote, unchallenged, the global warming and climate change thesis, largely on the basis of an increasingly-questioned 2006 seminar, rather than conduct any kind of public review of its climate change policy and how it came about, the BBC opted to fight off legal challenges and requests made to identify the scientists concerned under the UK's Freedom of Information Act.

Eventually, on 12 November 2012, climate blogger Maurizio Morabito published a list of those who attended the 2006 seminar that he had acquired through journalistic means.[34] In addition to 28 BBC staff, 30 specialists – presumably the 'best scientific experts'[35] referred to in the BBC report – attended. However, a cursory examination of the list of experts shows that only four could be classified as scientists (fewer, if post-normal scientists are classified differently to scientists). The rest of the attendees were predominantly climate campaigners, environmental activists, think-tank members, NGO staff and journalists – plus one member of the Church of England. A whole spectrum of institutional tools and personal skills necessary to advance the Climate Change cause and support a regime of climate truth was present, all reinforced by the backing of a global media outlet with a long-standing reputation for balanced neutrality. According to Morabito's published attendance list – which the BBC has not denied – every element of Michael Mann's recommendations for climate action to change the public's perception of the debate was present: scientists (though only a few), social scientists, creators of public policy, communications experts, sympathetic private financial foundations, suitably interested private sector institutions, mainstream media in the form of the BBC, campaign groups and Colin Challen, Member of Parliament and Chair of the All Party Group on Climate Change.[36] And perhaps most significantly – as confirmed by Richard North – no dissenting voices.

It appears that this is the point where the BBC's role in helping to create a regime of climate truth in support of the Climate Change cause was formalised. Yet the mere presence of a broadcaster, alongside numerous interested and sympathetic parties does not automatically equate to a change in public opinion or behaviour. Like any aspect of a military campaign, tactical battles in the climate truth wars need to be carefully planned with specific strategic outcomes in mind.

Climate facts, common sense and changing behaviour

The ultimate goal of any regime of truth is – as far as possible – to have its claims accepted as unquestioned and unquestionable orthodoxy to the extent that they cease to be challenged. In the West the Catholic Church managed such a feat for a millennium, using numerous institutional, social and legal levers. When things got desperate even a violent Inquisition was used to ensure that Catholic doctrine shaped how people should live their lives, express their beliefs and subordinate themselves to the authority of those who controlled that particular regime of truth in the name of the Christian God. However, much more subtle methods of shaping beliefs and behaviour are needed in the twenty-first century, as revealed in a report by the Institute for Public Policy Research (IPPR):

> the task of climate change agencies is not to persuade by rational argument but to develop a new 'common sense'...interested agencies now need to treat the argument as having been won, at least for popular communications. This means simply behaving as if climate change exists and is real, and that individual actions are effective. The 'facts' need to be treated as being taken-for-granted that they need not be spoken.[37]

The IPPR openly acknowledges that the purpose of the report is to show how behaviour and attitudes can be shaped in relation to climate change as: 'part of its project on how to stimulate climate-friendly behaviour in the UK.'[38] The IPPR describes itself as 'the UK's leading progressive think-tank.'[39] Like countless other thinktanks, it formulates policy ideas across a range of areas of interest and campaigns for their implementation through political lobbying and the shaping of behaviour patterns. In their own words: 'Our events programme brings high profile politicians and leading thinkers to a wide range of audiences.'[40] In all of this the IPPR is unexceptional: it is what thinktanks do across the whole spectrum of political outlooks from Washington to London to Berlin. They promote particular causes – usually with some anticipated social benefit – at the behest of their individual or corporate funders. The underlying ethic for such political activists is to see its interests – and by extension the interests of its supporters and funding streams – enacted as public policy. The truth is what they present it to be, in pursuit of the ends set down by their backers.

Lobbyists, policy research groups and thinktanks do not operate within an ethic of climate change anxiety or global warming prevention. Others like Greenpeace – a significant donor to the IPPR[41] – do that. The ethic of the lobbyist is purely based on the implementation of policy and the changing of behaviour, all in line with the aims of the funder. Tobacco companies employ lobbyists to represent their interests and advertisers to 'sell' their brand and encourage use of their products (no mean feat given the long-established, cancer-inducing properties of tobacco). Oil companies do likewise (though, like cigarette manufacturers, their product is hard to 'sell' as an ethical choice), and so do environmental organisations (a much easier 'sell'). If attitudes and behaviours are influenced and changed by lobbyists and publicists, the thinktanks have been successful, otherwise they have failed. The morality of the 'truth' being expounded, sold or otherwise promulgated is a separate matter entirely.

Ethically speaking, the work of social activists – and this applies to both advocates and opponents of AGW – is two steps removed from the ethical demands of the scientific method. Scientists are expected to conform to particular codes of practice within their respective fields (I will call this Ethic 1). Based on the evidence produced by scientists, ideologically-driven value judgements can be made about the social implications of their scientific findings (Ethic 2). Then finally, shaping behaviour as a consequence of the first two steps is a matter of policy implementation (Ethic 3). However, the Climate Change cause that dominates the climate truth wars in the twenty-first century has an inbuilt, potentially damaging flaw: it assumes that the three distinct ethics and their associated truth claims I have identified (science, ideology and policy implementation) are part of one overarching ethic that can be summed up under the worthy Climate Change cause of saving the planet.

Perhaps not surprisingly, the approach of the climate scientist-activist has had unforeseen but potentially problematic consequences. By venturing beyond the relatively safe territory of the scientific method and the qualified claims that it frequently produces and into the realm of political persuasion, Climate Change adherents potentially undermine the very regime of truth that they are trying to sustain. Those who have taken Schneider's advice literally – 'we have to offer up scary scenarios, make simplified, dramatic statements, and make little mention of any doubts we might have'[42] – are the greatest culprits. Consider the implications of these three elements: 'scary scenarios', 'simplified, dramatic statements', and 'little mention of any doubts'.

Scary scenarios have been used since time immemorial to frighten, cajole or persuade people to behave in particular ways – often with religious motivation to encourage some kind of ethical standard. I have very clear memories of one particular doomsayer from my youth who would regularly walk the High Street of my university town wearing a sandwich board that said on the front: 'THE END OF THE WORLD IS NIGH'. On the back it offered concerned citizens a choice: 'REPENT OR DIE'. I do not know who the man was but when you have seen the same warning week after week and year after year, without the world coming to an end, his ongoing articulation of the imminent threat was the very thing that undermined it most of all. Doubt and fear gave way to ridicule. It is easy now to mock the Y2K panic that gripped the world on the eve of the twenty-first century, when governments, banks, armed forces and countless other organisations planned for catastrophic computer melt-down. It did not happen. By coincidence, as I make a first draft of these notes it is 7.52pm on Thursday, 20 December 2012. According to an ancient Mayan prophecy the world will end tomorrow. So, to be on the safe side I am going to spend the evening with my family and resume on Saturday if I am still here. ... [43]

And so Saturday arrives. That is the problem with Armageddon – if you promise it you need to deliver. For many Climate Change advocates and environmental activists the 2004 film *The Day After Tomorrow* was an unqualified success: it reached a massive global audience and its message about taking care of the planet by acting swiftly to combat climate change could have been written by the IPCC. For more subtle, cultural reasons I would suggest it is well on the way to being an unmitigated disaster for those – including myself – who are concerned about the long-term environmental well-being of our planet, principally because it reinforced a short-termist mindset by presenting a fictionalised account of how quickly climate disaster might happen.[44] In less than two hours in a darkened cinema – and in only two days as it was presented on screen – the world went from stable to utter turmoil: millions, possibly billions, dead and familiar geo-political structures destroyed.

It is feasible that the IPCC's 2007 projected global mean temperature rise of up to 6.4°C[45] in the twenty-first century will come to fruition, along with the multiple disastrous consequences that have been associated with it: polar ice melt, glacial melt, rising sea temperatures, colossal coastal flooding, inland drought, crop failure and so on. It is difficult, however, to sustain the level of personal and collective apoca-lyptic anxiety necessary to prompt individual and societal action when, at the time of writing this in 2014, the mean global temperature has not risen since 1997. Further, the final section of the chapter highlights the vulnerability of the regime of Climate Change truth as it is currently constructed: a vulnerability that goes back to the relationship between science and post-normal science.

Theory, action and scientific credibility

Kathryn Humphrey, from the UK's Adaption to Climate Change programme, wrote to Phil Jones in 2009 to discuss the benefits of a Government funded Weather Generator, prior to a meeting with Hilary Benn, then Secretary of State for Environment, Food and Rural Affairs. Humphrey was concerned that Ministers were worried about criticism of the project by other scientists:

> I can't overstate the HUGE amount of political interest in the project as a message that the Government can give on climate change to help them tell their story. They want the story to be a very strong one and don't want to be made to look foolish. Therefore, every time they hear about any criticisms from anyone, they jump.[46]

We see here an emphasis on the Government telling its climate change 'story' – making the public aware of the risks of global warming with the aim of shaping behaviour by means of 'Green' policies, taxes and incen-tives: more recycling, less waste, willingness to pay more for renewable energy, and so on. Governments prefer to use uncomplicated 'stories' or messages when they attempt to shape the attitudes and behaviour of the people they govern: the more democratic the government system the more subtle and pervasive the messages need to be as truth is brought to bear in the exercise of power. Crucially, such stories need to be seen as credible, reliable and able to stand up to rigorous scrutiny. The scientific method can support these requirements. Scientists question, it is what their moral code requires them to do. However, as I argued in Chapter 1, providing a strong story usually means dropping the caveats and cautions that surround credible and reliable scientific findings. This leads to scientific truth being further subordinated to the political dimension of Climate Change. Regardless of the level of political support the regime of climate truth may enjoy, it is at its most vulnerable where claims are overstated and the science is not robust enough to support them.

Unverified scientific claims damage the Climate Change cause – or any other cause – when their findings fail rigorous verification (I will not spend time here 'debunking' ill-informed or 'big oil'-funded denialist propaganda that I consider to be compromised in the first place). The most high profile example in recent years involved Rajendra Pachauri, Chair of the IPCC in 2007 when it issued a report warning about the severe threat global warming posed to Himalayan glaciers. The report stated: 'Glaciers in the Himalaya are receding faster than in any other part of the world ... and, if the present rate continues, the likelihood of them disappearing by the year 2035 and perhaps sooner is very high if the Earth keeps warming at the current rate.'[47]

This was a hugely significant claim, given the millions of people spread over several countries on that continent who depend upon glacier melt to feed the rivers that support their lives and livelihoods. The IPCC's claim was subsequently refuted by Dr Vijay Raina, an Indian Government-appointed glaciologist, in a report entitled 'Himalayan Glaciers – A State-of-Art Review of Glacial Studies, Glacial Retreat and Climate Change'.[48] Raina's study examined more than 20 glaciers, identifying unique behaviour in all of them: some were shrinking, some were growing, and some had remained a constant size for decades. The study concluded that it would be premature to say that global warming was causing unusual glacial retreat.[49]

The level of publicity achieved, and the degree of anxiety caused, by the 2007 IPCC report ensured that when Raina's findings were made public, they would have both national and international impact. Rajendra Pachauri – still Chair of the IPCC when Raina's report was released – was interviewed on Indian television and asked about Raina's research and its clear challenge to the IPCC's published position on the severity of global warming-induced Himalayan glacier melt. Pachauri chose not to engage with Raina's work at the level of objective science: no immediate detailed analysis of results and no attempt to replicate the findings. He did, however, reject the study sample as too small to be of statistical significance.

Tellingly, Pachauri responded instead at a subjective level, criticising Raina and questioning both his personal integrity and scientific judgement. Despite not having read the critical report, Pachauri felt able to respond in the boldest of terms to his TV interviewer: 'These findings are totally wrong.'[50] No hesitation, no equivocation. With complete confidence he declared that the 2007 IPCC findings on Himalayan glacier melt were based on 'peer reviewed literature' from 'prestigious, credible journals.' In contrast, he said of Raina's work: 'This is, if I may say so, voodoo science. This is not science.'[51] A measured, science-based refutation of the arguments was rejected in favour of derogatory grand-standing that bordered on an attempted public character assassination. The climate truth wars got deeply personal. Pachauri sought to represent Raina as unethical, wilfully promulgating findings that were misleading, wrong and unscientific. The most telling aspect of his response occurred when he opted for a political justification of the IPCC's position:

> I don't think he has any business questioning a body that has estab-lished its credentials over the last 21 years, and whose reports are accepted by every government of the world...I question these find-ings completely, they don't make sense to me at all.[52]

Pachauri's response only makes sense if viewed in the context of a regime of climate truth and the Climate Change cause that he was defending. Pachauri sought to validate the IPCC research by pointing to the number of governments around the world who accept it. In other words, the scientific argument was rejected in favour of the political as truth was wielded in an exercise of power. However, his was a circular argument given that it was the governments of the world that originally established the IPCC to keep them informed about the dangers of global warming. Finally, Pachauri classed Raina as a denialist who rejects the man-made global warming thesis. A cursory glance at Raina's summary of findings would have told Pachauri that his critic was no denialist. Raina's claims were narrow, specific, research-based, non-political and even acknowl-edged that global warming could and would influence glacial melt.

Despite Pachauri's politicised denunciation of Raina's work, deliv-ered with the full institutional authority that he wielded as Chair of the IPCC, the public nature of the disagreement required that the science had to be similarly refuted. Unfortunately for Pachauri's personal repu-tation and the almost unchallenged and unchallengeable authority of the IPCC on matters concerning global warming and climate change, Raina's research withstood peer scientific scrutiny: that is, scientific in the 'normal' science sense. Pachauri's defence of the IPCC's claims about Himalayan glacier melt ultimately failed – publicly and embarrassingly – on scientific grounds. As a result, the IPCC was forced to issue a state-ment retracting the claims made in its 2007 report:

> It has, however, recently come to our attention that a paragraph in the 938-page Working Group II contribution to the underlying assess-ment refers to poorly substantiated estimates of rate of recession and

date for the disappearance of Himalayan glaciers. In drafting the paragraph in question, the clear and well-established standards of evidence, required by the IPCC procedures, were not applied properly. The Chair, Vice-Chairs and Co-chairs of the IPCC regret the poor application of well established IPCC procedures in this instance.[53]

The retraction itself was worded in such a way as to limit damage to the IPCC and the regime of climate truth that it has so carefully helped to build and so staunchly defends. It suggests – though there is no way of demonstrating whether the claim is right or wrong – that there is only one erroneous paragraph in a report that is 938 pages long. The Chair (Pachauri) expressed regret but there was no personal or institutional apology for the aggressive, outspoken, deeply personal criticisms of Raina made on Indian television. Furthermore, Pachauri did not withdraw his description of Raina as a denialist. In the climate truth wars, being scientifically correct at all times is not the priority: maintaining the political and ideological 'truth' about the cause is more important.

Crucially, this unedifying incident highlights the key vulnerability of Climate Change and its regime of climate truth. The post-normal science-based approach that is advocated by many climate scientists can be undermined by careful, focused application of the strict, established codes of normal science to specific, narrow examples: thereby weakening the case as a whole. Under duress, the accuracy of Raina's research outweighed the constituted truth of the Climate Change cause as presented by Rajendra Pachauri. It was, however, only one small tactical victory in the truth wars, but it has demonstrated to other opponents of the regime of Climate Change truth (many of whom, like Raina, accept AGW) how it might be undermined. Consequently, Chapter 3 will analyse how the imprecision of aspects of the overall scientific case for global warming (no matter how robust individual studies are) weakens the case as a whole. The fact that nobody in the 1990s shouted from the roof-tops that the world should shortly expect a near 20-year hiatus in global mean temperature rises leaves scope for politicians to reshape their priorities and relegate the significance of Climate Change truth and the importance of global warming as an impending threat. One potential moral hazard prompted by Climate Change advocates is this: that short-term policy advancements based on over-stating risk (for even the best of intentions) eventually gives way to long-term disillusionment and disengagement by the public and politicians alike.

3
One World, Two Visions

In the opening decade of the twenty-first century, Climate Change and its regime of truth appeared to hold an unassailable dominance in the climate truth wars, with sceptical opponents effectively side-lined – though not silenced – and a global framework in place to tackle the threat of global warming. On the scientific side, the IPCC had produced several assessments which, in turn, had been accepted by the governments – most of the governments in the world – that had commissioned the reports. The interrelatedness of power and climate truth was clearly identifiable as numerous Green policies were enacted, especially in the most advanced economies, with a view to at least slowing down the increases in carbon released into the atmosphere. The geopolitical response that began with the 1997 Kyoto Protocol, which saw 37 industrialised nations plus the European Union committing to large cuts in greenhouse gas emissions (GHG), continued through the 2001 Marrakesh Accords, and led most recently to the 2012 Doha Amendment of the Kyoto Protocol.[1] Conclusive victory for Climate Change advocates would appear complete: except that it is not.

Three factors threaten the Climate Change cause and its dominant regime of truth: two external and one internal. First, opponents continue to challenge aspects of the bold post-normal science-based claims on which the regime stands, encouraged by unforeseen events – no matter how small – like the 2013 Arctic sea-ice figures. Data from Europe's Cryosat spacecraft showed a 50 per cent increase in Arctic sea-ice volume at the end of the melting season compared with the record low of 2012,[2] which had prompted talk of opening up the Northwest Passage to enable ships to sail through what had previously been areas of dense pack-ice. More damaging to the Climate Change cause were events in December 2013 and January 2014, when the Australasian Antarctic

Expedition embarked upon a high profile research expedition to demonstrate that the East Antarctic ice sheet is melting as a result of global warming. Unfortunately for the project the Russian ice-reinforced ship it commissioned, *MV Akademik Shokalskiy*, got trapped in dense pack-ice in Commonwealth Bay with dozens of global warming scientists, journalists and other climate activists aboard. Worse, the ice was so dense that two rescue ice-breakers repeatedly failed to reach and free the *MV Akademik Shokalskiy*. Consequently, the researchers had to be airlifted to safety in the glare of global publicity. Compounding the embarrassment of the research team and damaging public perceptions of global warming in the process, that area had not been ice-bound 100 years earlier when Antarctic explorer Douglas Mawson conducted his expedition: a point that was highlighted in mocking tones by newspaper editorials around the world.[3] In the climate truth wars, scientific nuance provides little protection from public mockery and negative perception. These two incidents involving Arctic and Antarctic ice have to be seen in the context of several decades of climate science and 30 years of declining Arctic sea-ice levels. However, overstated Climate Change certainties leave the regime of climate truth vulnerable to unexpected or inexplicable events.

The second threat to the regime of Climate Change truth lies in the ongoing global mean temperature pause, which puts increasing pressure on climate modellers to explain why it came about and how long it will last. Given the emphasis placed on measured increases in global temperatures from the 1970s to the 1990s – the global warming that would lead to Climate Change – it is a point of vulnerability that can be counterattacked by opponents or potential supporters who are not convinced of every part of the regime of climate truth. When the IPCC Working Group AR5 Summary for Policymakers was released on 31 March 2014, further opportunities for criticism were provided to Climate Change opponents when senior IPCC figure Chris Field stated, "There is no pause," reinforcing Michel Jarraud's similar statement, 'I really refuse to accept that we can talk about a pause.'[4] Lorraine Whitmarsh has researched popular understanding of the terms climate change and global warming and found that the latter expression caused greater public concern than the former.[5] If global warming is publicly perceived to have stopped, any climate threat may be seen by opponents or climate agnostics (those who are open to being convinced but currently might not be) to have diminished or gone away. Subtle arguments about the sea absorbing and hiding heat increases do not make appealing or scary headlines of the kind on which the regime of truth was initially constructed. Consequently, it will become much more difficult, impossible perhaps, to persuade the populations of developed and developing countries to pay the vast costs of de-carbonising their economies.

The third threat is a civil war within the Climate Change cause to shape the world's political response to climate change. This internal battle is rooted in the contradictory ethical reasons that individuals have for accepting and supporting the regime of climate truth in the first place. The result is an ideological fault line which poses as much of a threat to Climate Change as the normal science/post-normal science contradiction it also contains, and is summed up in the following ethical choice. What is given moral priority by Climate Change advocates: care of the environment or quality of human life? This question is significant because in the absence of a definite 'right' way, those who wield political power at national levels have the freedom to invoke whatever answer to this question best serves their purposes. The 'truth' about the best response to climate change – assuming a specific response is still needed – is that which best suits the policy makers, their ideological motivations, and the political constraints within which they operate.

Of course, the interests of humans and the environment are interconnected and politicians will claim that both can be achieved simultaneously. However, one is always given a higher priority than the other. Two examples of opposing priorities are examined below, the first from religious, Christian discourse and the other from secular, ecological discourse. Consequently, as the chapter will go on to discuss, prioritising the environment over human concerns, or vice versa, has a significant impact on political responses to Climate Change. Furthermore, I will argue that this ideological fault line is at the heart of recent, major political climate-related decisions such as the withdrawal of several countries from the Kyoto Protocol and the unwillingness of leaders of major established and emerging economies to take climate action that will impede economic recovery and growth. In the politics of truth, the global response to Climate Change – regardless of the findings of the latest IPCC report and its continual upgrading of climate threats, even if they happen to be 100 per cent accurate – has been relegated behind matters of national economic self-interest.

The ethical basis of climate concern

In 1976 my family moved from Scotland to Malaysia and in the maelstrom of new sights, sounds and smells, one in particular still stands out in my memory – a small group of shaven-headed Buddhist monks in bright orange robes slowly walking down a chaotic street. They moved calmly and deliberately, as though disconnected from the tumult round about them, each sweeping a soft broom across the ground in front of him prior to every careful step. The purpose of such care was not to clean the streets and pathways of Penang but to ensure that they did not kill any ants or other crawling insects by inadvertently stepping on them. The moral duty of the monks, as they expressed their religious piety in this way, was to preserve the lives of all living creatures, either directly or indirectly. Direct killing was avoided by sweeping where they would place their feet, while indirect killing was avoided through a vegetarian diet and by not wearing shoes or clothes made from animal products. In many ways it was an ideal and idealised environmentally friendly lifestyle.

However, even as a ten year old I understood that the monks' day-to-day existence was highly self-disciplined, slow-moving, inconvenient and, frankly, not much fun. Such self-sacrificial asceticism was considered extreme even by my Buddhist school friends and definitely not as appealing as Adidas leather football boots or Nasi Goreng Ayam.[6] Their existence was the embodied rejection of self-interest. Self-interest manifests itself in many ways: a desire for material products of every kind (as far as budgets permit); driving a bigger car with a bigger carbon footprint and driving it quickly to save my valuable time; visiting far-flung parts of the world simply because it seems appealing; or voting for the government most likely to maintain my preferred lifestyle (assuming you have a vote). Moral duty and self-interest cannot simply be thrown together on a whim in the hope of producing a coherent framework for living. Contrast, if you will, aspects of moral duty to the natural world demonstrated by the Buddhist monks with that set out by the Roman Catholic Church.

In his inaugural address in 2013, Pope Francis I stood before hundreds of thousands of the Roman Catholic faithful in St Peter's Square and urged Christians to protect nature and the beauty of the created world, 'respecting each of God's creatures and respecting the environment in which we live.'[7] He continued: 'Please, I would like to ask all those who have positions of responsibility in economic, political and social life, and all men and women of goodwill: let us be "protectors" of creation, protectors of God's plan inscribed in nature, protectors of one another and of the environment.'[8] These few words capture not only Francis's practical aims but the theological principles on which they are based. He spoke of 'the created world,' 'God's creatures,' 'God's plan inscribed in nature.' Nature and the environment are understood in Catholic doctrine, as they are similarly understood in rival Christian doctrines, as part of Divine creation. The very planet we live on is understood to be nothing less than an expression of the creative will of God. However, while protecting nature may be considered part of the Christian's duty and day-to-day existence, historically speaking it has rarely been a priority in the way that it is for the Buddhist monks I encountered.

In 1967 Lynn White wrote of the relationship between the Christian and the environment: 'Especially in its Western form, Christianity is the most anthropocentric[9] religion the world has seen.'[10] White went on to argue: 'Christianity ... not only established a dualism of man and nature but also insisted that it is God's will that man exploit nature for his proper ends.'[11] In this view, far from being environmentally concerned, Christianity prioritises humans over nature more than any other religion in the world. In the twenty-first century there remains considerable evidence to support this view.

The ongoing industrial and material success of the United States suggests links between Christianity, an advanced and voracious consumer culture, aggressive capitalism, and more greenhouse gas emissions per head than any other country in the world.[12] These relationships are formed in a country where 163 million Americans are affiliated to more than 200 national church bodies.[13] It takes a huge amount of 'stuff' to maintain the lifestyle of the average American. That level of production, consumption and pollution – carbon emissions and other damage to the environment – can only co-exist with such widespread adherence to Christianity if, generally, the majority Christian worldview prioritises the success and affluence of the individual over care for the environment in which we live and the preservation of natural resources. (Europe also has high levels of production, consumption and pollution but is rapidly becoming de-Christianised.) The contrast with the Buddhist monks and their desire to preserve the ants under their feet does not really need to be drawn any further. As we will see later however, this contrast in outlook provides a major front in the climate truth wars and the political response to global warming.

Lynn White had not set out simply to apportion blame for what he termed our 'ecologic crisis.'[14] He also offered an alternative Christian basis for a more environmentally aware and ecologically sympathetic engagement with the world based on the life, teaching and example of Saint Francis of Assisi: the Catholic friar who lived eight centuries ago. Francis turned his back on the wealth and mercantilism of his father, rejected materialism and opted instead for a poor, humble existence *alongside* nature rather than *dominating* nature. In 1979, Pope John Paul II spared no hyperbole when he declared Francis of Assisi 'the heavenly Patron of those who promote ecology.'[15] He continued: 'The poor man of Assisi gives us striking witness that when we are at peace with God we are better able to devote ourselves to building up that peace with all creation.'[16] This is the same Francis of Assisi who so inspired Cardinal Jorge Bergoglio that he adopted the name Pope Francis upon his election in 2013. Note, however, the priorities in John Paul's words: peace with God comes first, followed by peace with creation. He also added: 'In the words of the Second Vatican Council, "God destined the earth and all it contains for the use of every individual and all peoples".'[17] The official position of the Roman Catholic Church stresses that the Divine intention for the Earth and its resources is to serve humanity, to be subordinated to 'every individual and all people' for their use.

So we see within Christianity considerable tensions and divisions over where the environment, the *created order* to put it in religious terms, should fit into a Christian's priorities: from the 'prosperity' gospel preached by extremely wealthy pastors from Dallas to Lagos, to the resurgence of neo-Celtic Christianity in the UK and beyond since the 1990s. In the former, exemplified by the huge material success of preachers like Joel Osteen, we find the following: 'If you foster an image of defeat and failure, then you're going to live that kind of life. But if you develop an image of victory, success, health, abundance, joy, peace, and happiness, nothing on earth will be able to hold those things from you.'[18] In the latter we find a desire to pursue a non-materialistic co-existence with nature and the rest of humanity. It should not be hard to guess which approach to Christianity is thriving around the world, and which is a niche interest. The contradiction within these two Christian perspectives on life and humanity's relationship with the environment is vast, even supposing an authoritative interpretation of 'the will of God' could be found – a notion that will be challenged by other religions and rejected as preposterous by atheists and humanists alike.

Writing from a Buddhist perspective, Stephanie Kaza – an academic and environmental campaigner since the 1960s – provides yet another view on the relationship between humans and the environment. Her starting point is the notion of reducing harm. More specifically it is about reducing harm that we do to the planet and living in greater harmony with nature. Beginning with the American meat-based food cycle she describes the ways in which the environment is damaged by intensive farming and related resource depletion, followed by consideration for the harm to animals and plant life. Then finally she describes the harms – heart disease and obesity – that humans do to themselves through their food production and consumption habits. She observes: 'Reducing harm to ourselves is a viable and important aspect of reducing environmental impact, reflecting the recognition that we too are part of the environment that is under siege.'[19] Referring to the 'interdependent nature of reality,'[20] Kaza goes on to demonstrate what for her is the interdependent reality of nature. In terms of moral priority, however, welfare of the environment and its constituent, interconnected parts outranks concern for the people whose existence would do it harm.

For some, all of this religious talk might be as relevant as medieval debates about how many angels can dance on a pinhead. However, an overwhelming proportion of the world's population holds some kind of religious belief or affiliation, which in turn plays a part in how individuals determine the truth about climate change, global warming, and the environment. Moreover, as we will see later, those subjective beliefs and motivations play a crucial role in deciding whether or not action to prevent global warming should be taken and if so what should be done. Not only that, individual beliefs also inform what we are trying to achieve through environmental engagement – eternity in heaven, a better future earth, a socialist equality for all, a better standard of living, and many more.

An alternative, non-religious ethical basis for adopting Climate Change and supporting the regime of climate truth can be found in the ecological movement. For many, Earth and its natural environment is an object of veneration, with humanity providing the greatest challenge to its ongoing health and existence. Worship of the earth, sky, sun, seas and other natural phenomena has occurred from the earliest human existence, as witnessed by cave paintings and archaeological evidence all over the world. However, in recent decades one strand of the ecology movement has combined scientific understanding, environmental awareness and devotion to the biosphere in framing an ethic for the global warming age.

In 1970 Edward Goldsmith founded *The Ecologist*, an environment-focused journal that would go on to achieve global standing and influence. In the first edition he set out his personal views, his political priorities and the editorial tone of the journal, describing the relationship between humans and Planet Earth as that of a parasite to a host, a 'disease [that] has spread exponentially.'[21] In the relationship between humans and the biosphere we occupy, the latter is granted priority, with human beings merely seen as a problem to be managed in order to protect the planet as far as possible. Consequently, Goldsmith advocated radical solutions to the 'problem' of humanity's relationship with the environment: constraining personal liberty, cutting economic growth, and curbing 'the march of progress.'[22] Further, the global population should be 'planned,' as should consumption levels and standards of living;[23] political ideology and environmentalism neatly synthesised.

Goldsmith also proposed an early shift away from the constraints of normal science, saying: 'Science is too serious a matter to be left to the scientists.'[24] He also described how to best shape the behaviour of the parasite (humans) that is leeching from its host (the environment):[25] by using national educational institutions to inculcate pro-environmental values in individuals so they could learn how to live properly in relation to the eco-system that sustains all human life.[26] Goldsmith's blurring of science, Marxist-inspired ideology and social action, is designed to ensure that people conduct themselves correctly – as he sees it – in relation to the environment.

Many aspects of Goldsmith's vision have been appropriated by the Climate Change cause and its regime of truth. His theme was echoed in 1991 by the Club of Rome, an ecologically-focused, ideologically-inspired collective of influential individuals who work together to promote the environment and policies on sustainability:

> In searching for a common enemy against whom we can unite, we came up with the idea that pollution, the threat of global warming, water shortages, famine and the like, would fit the bill. In their totality and their interactions these phenomena do constitute a common threat which must be confronted by everyone together. All these dangers are caused by *human* intervention in natural processes, and it is only through changed attitudes and behaviour that they can be overcome. The real enemy then is humanity itself.[27]

When this statement is set alongside the sentiments of Pope John Paul II in 1990 there are some similarities. We see concern about unrestrained materialism and care of the environment set alongside the need to protect the poorest and most vulnerable people from starvation and drought: all of which require new attitudes and behaviours. A more holistic co-existence with nature is advocated by all: secular and religious, theist and atheist.

More problematically, all of these similarities must be viewed in light of the fundamental difference in attitudes to the relationship between humanity and nature. On the one hand the Edward Goldsmith-Club of Rome-Buddhist approach clearly prioritizes ecology to the extent that humanity is an enemy to be controlled, subjugated, defeated and – *in extremis* – eradicated like disease, in pursuit of the environmental cause. Further, in secular ecology discourses, the ethical individual does not pursue some heavenly reward in the afterlife, instead preferring care for the environment in this life. On the other hand, the Catholic-Christian-Western materialist perspective seeks the advancement of humanity's quality of life – across a spectrum that ranges from basic survival needs to conspicuous consumption – and will tolerate or even encourage environmental harm in the process. The modern, previously Christianised West shares little of the religious commitment that shaped its history, especially its pre-Enlightenment history. However, echoes of that distant past still remain – usually unrecognised and unacknowledged – shaping attitudes, priorities and behaviour, now exported beyond the Western world through the globalisation of materialism.

Green politics

After three decades of careful planning, scientific research, ideological advancement and institutional acceptance, the Climate Change cause had reached such a dominant position in national and international global warming and environmental discourse that in 2006, British Prime Minister Tony Blair was able to say to the Conference on Climate Change and Governance in Wellington, New Zealand:

> I think in terms of the long-term future there is no issue that is more important than climate change ... Climate change is not just an environmental challenge, but is a threat to the global economy and to global security. The projected rise in the global average temperature will have catastrophic effects, increasing the frequency of extreme weather events (floods, cyclones and hurricanes). Tackling climate change requires a sustained global effort to reduce emissions of greenhouse gases through cleaner energy, transport and changes in technology and behaviour.[28]

Blair demonstrated no doubts, offered no equivocation and added no science-based cautions or caveats. The vision of Stephen Schneider, Edward Goldsmith, the Club of Rome, Michael Mann and countless others had apparently come to fruition as global warming and Climate Change took centre stage in not only environmental policy but in economic, energy, transport, technology and security policy as well. At the risk of being embarrassed by future events, I will go so far as to say that the regime of climate truth reached the point of its furthest advancement in the climate truth wars around this time. The need to ameliorate the catastrophic threats posed by global warming had reached the political mainstream in almost every country and had been accepted by political leaders on the world stage. The need for individual behavioural change was accepted almost universally: humans are destroying the planet on the way to destroying themselves. To qualify that statement – it might be more accurate to say that almost everyone agreed that everyone else had to change their behaviour. The following year, illustrating its importance to Blair's government, the UK prompted a UN Security Council debate on the impact of climate change. Jan Kubis, Minister for Foreign Affairs of Slovakia, was an early contributor to the debate who highlighted that 'there was now an effective consensus among the world's leading scientists that there was a discernible human influence on the climate and a link between the concentration of carbon dioxide and the increase in temperature [and that it] was time to consider the policy dimensions of climate change.'[29] The policies advocated by Kubis included: emissions trading as a means of reducing carbon emissions; investment in low-carbon technology; and 'nuclear energy as a cleaner choice.'[30] Almost immediately, the complexity of the issues and interests involved was highlighted with an African perspective. L. K. Christian, the Ghanaian representative, hoped that 'repeated alarm' about climate change would result in 'action that is timely, concerted and sustainable.'[31] Such action, however, had a distinct and specific political-social dimension: 'the issue of climate change in Africa should be framed in terms of how to combat the phenomenon without compromising the targeted 8 per cent growth rate needed to reduce poverty.'[32] From Christian's comments it is clear that in terms of political goals, poverty reduction was prioritised over the environment and climate change. From yet another perspective, Robert G. Aisi, from Papua New Guinea, described the climate change concerns of the Pacific Islands forum in terms of a threat to their very existence: mass population dislocation and a breakdown of social cohesion and individual and collective identity. He equated climate threat to their small island nations to the 'dangers guns and bombs posed to large nations.'[33]

This snapshot of concerns raised at the UN highlights significant political differences in the approach to Climate Change: from an economic and technical challenge in Europe, to subsuming climate change within the higher goal of eradicating poverty in Africa, to maintaining the existence of Pacific islands, their sovereignty, culture and social identity. All of which is predicated on the regime of climate truth and its central claim: carbon dioxide and other greenhouse gases emitted through human activity are driving up global temperature at a dangerous rate which will, in turn, melt polar ice caps and cause global sea level rises. At the same time, climate change will impact upon global weather patterns to such an extent that crop yields will struggle to sustain the world's population. Hence – so the argument goes – the need for extreme, costly and collective political solutions, such as the Kyoto Protocol and its successors.

What becomes immediately clear in revisiting the 2007 UN debate on climate change and many related political statements of that time, is the consistency of Climate Change *realpolitik* at work: every speaker spoke of the global consequences of climate change but framed their contribution in terms of their own country's or region's political interests. As a result, claims to climate *truth* – when expressed in self-interested political terms – become essentially meaningless as a concept with universal application. And if this is the case there can be no objective, generalised and accepted morality to which individuals can conform amidst the political and ideological contradictions and inconsistencies in Climate Change discourse (even setting aside the pivotal question of the reliability of post-normal science-based climate models and their projected outcomes).

With no *objective* means of measuring or specifying climate-related morality and political choices, the only way Climate Change can now be reasonably understood is as a domain of *subjective* moral experience. Hence, individuals are encouraged to form themselves as ethical by accepting the purported certainties of post-normal science, ideologically-based environmental claims and political responses. That reliance on subjective experience, however, poses the biggest threat to the regime of truth that has advanced so successfully to dominate the climate truth wars since the 1970s. Any chink in the armour of climate certainty – and the ongoing, unexplained pause in temperature rises is such a weakness – offers politicians the opportunity to downgrade the climate threat, renege on climate spending commitments made during the economic boom years of the 2000s, and reorder their priorities. This reprioritisation has provided the biggest setback yet to the Climate Change cause since the global recession began in 2007, accelerated through the banking collapses of 2008, and continues to destabilise large parts of the global economy.

The winds of change

The UK's 2008 Climate Change Act, like similar climate laws in other developed countries, was introduced at the height of the financial bubble which was about to burst as part of the global financial crisis. Environmental aspirations in the early 2000s were set out at a time of at least perceived national and individual wealth, with financial implications for the country and for the consumer that would become increasingly unpalatable over the next few years. The rule of renewable technologies seemed inevitable: solar power and wind power would provide the foundation of a carbon-free future, their exorbitant costs paid willingly by a population largely persuaded by Climate Change ideology and its truth claims. Throw in the dire predictions that Peak Oil happened sometime in the 2000s, resulting in the need for alternative, non-carbon fuel sources in the short, medium and long terms,[34] and Richard Heinberg's warning that oil from shale was but a 'hollow promise,'[35] there seemed to be no alternative. As recently as 2011, the British Government sought to build on the radical Climate Change Act of 2008[36] – which looked to legally enshrine environmental protection as a priority within government policy – by setting out its vision for an environmentally friendly future. It states: 'We want the EU to become the world's largest green economy and market for environmentally sustainable goods and services. We will work with our partners to put in place appropriate strategies and sectoral policies, to achieve low-carbon, resource-efficient growth.'[37]

At this point in recent history the environment was the stated ethical and policy priority in the climate truth wars: the political authorities were persuaded and key institutions were acting upon that persuasion. From a Climate Change perspective, power and truth were operating in tandem for the benefit of the environment. Growth in individual standards of living would be allowed to suffer *slightly* in the rich West for the greater good of the environment. Anyway, with huge and growing demand for dwindling oil supplies, and soaring oil and commodity prices, numerous governments argued – and many people agreed – that there was no option. Between 2008 and 2011 the UK and Germany went further than any other nations in supporting the ideological prioritisation of the environment over out-and-out economic growth and individual wealth. Complicating matters, after the Fukushima disaster in Japan, Germany made the decision to phase out its existing nuclear power plants and cancel plans to build more in the future.[38] The use of renewable solar and wind power would be expanded at vast cost to the German taxpayer and German industry as the regime of climate truth and its environmental priorities shaped economic, industrial and energy policy. At least temporarily. Simultaneously, the UK was also putting the environment first when the Secretary of State for Environment, Food and Rural Affairs', Caroline Spelman announced that the UK 'places the value of nature at the centre of the choices our nation must make: to enhance our environment, economic growth and personal well-being.'[39]

In the United States, President Obama was sending out a similar message. In his 2010 State of the Union Address he spoke with typical impassioned eloquence about the existence and scale of climate change: 'I know that there are those who disagree with the overwhelming scientific evidence on climate change.'[40] The Climate Change cause now extended to the White House. By the start of 2013 Obama's commitment seemed as strong as ever:

> But for the sake of our children and our future, we must do more to combat climate change. Now, it's true that no single event makes a trend. But the fact is the 12 hottest years on record have all come in the last 15. Heatwaves, droughts, wildfires, floods – all are now more frequent and more intense. We can choose to believe that super-storm Sandy, and the most severe drought in decades, and the worst wildfires some states have ever seen were all just a freak coincidence. Or we can choose to believe in the overwhelming judgment of science – and act before it's too late.[41]

Despite the repeated message by climate scientists over the years that climate is not weather – usually to counteract claims that severe snowstorms made a mockery of global warming – Obama was quite content to use a number of severe weather events to advance the climate cause. Further, he was happy to repeat an established global warming truth when he pointed out that twelve out of the previous fifteen years had been some of the hottest years on record. However, an alternative climate truth – using the same evidence – preferred by opponents of Climate Change would say: 'Global temperatures have not risen for fifteen years and the overwhelming judgement of the climate science community is that they do not know exactly why that is, especially given that carbon dioxide levels in the atmosphere have risen consistently over the same period. In addition, climate scientists cannot *prove* exactly why that should be so. Also, while super-storm Sandy was highly destructive we can be grateful that the number of hurricanes this year is down on the seasonal average.' Truth is a malleable thing.

By the end of 2012, Prime Minister David Cameron was being accused of backtracking on his commitment to the environment. When challenged before the UK Parliament's Liaison Committee whether or not he had abandoned the principle that 'the environment is the foundation of sustained economic growth,' Cameron obfuscated. [42] On the one hand he stated categorically, 'This is a very green Government,' while going on to speak of 'wrestling with huge problems of debt, deficit, economic growth and all the rest of it.'[43] He also went further in distancing himself from the Green ideology he had previously advocated but which would limit his political room to manoeuvre as he sought British economic recovery in the aftermath of the global financial crisis:

> The debate I would have right now with the some in the green movement is that some in the green movement really want us to rule out gas, effectively, in a meaningful way. They want us to opt right now for nuclear, plus renewables, plus energy efficiency. Zip-that's it. I think that would be a mistake.[44]

In stark terms, Cameron and others – though not all – in his coalition government began to reorder the climate message. Increasing household and business energy costs that might more gladly have been borne during the economic expansion a few years earlier, were by 2013 becoming an increasingly poisoned political chalice. With the exploitation of shale gas taking off at an incredible rate in the US, and major emerging economies like China and India prioritising economic growth and individual prosperity, environmental idealism had resulted in the UK and Europe lagging behind their major competitors in terms of energy costs.

Events in Germany – where increases in industrial output continued despite the global financial disaster that laid low many economies around the world from 2007 onwards – are following a similar trajectory of environmental contradiction. Even the German economy is not immune to the backwash of global recession and the financial pressures of bailing out several failing European economies. In that context, Germany has been unable – or at least politically unwilling – to afford to continue the high levels of subsidies that it has been paying to underwrite the expansion of renewable energies – solar and wind power in particular. Concerns about standards of living have gradually taken priority over care for the environment. In January 2013, German Environment Minister Peter Altmaier announced that subsidies for renewable energy would be cut in order to halt the rapid rise in electricity prices. He stated: 'It is not acceptable that electricity consumers should keep bearing all the risks of the future costs [of renewable energy] on their own.'[45] Leading up to the September 2013 general election in Germany, Altmaier and the government of Chancellor Angela Merkel began to move away from their previously lavish ambitions for the future of renewable energy use in their country. However, as part of their cost-saving measures, financial exemptions for power-intensive industries were also to be reduced. They therefore upset both the environmental lobby and the industry lobby: politically-driven concerns for individual consumers (and their soon-to-be-needed votes) were placed above environmental interests and even industrial competitiveness.

In another development, Germany's opposition to the widespread use of fracking (as seen in the US) left it just one energy option in the short-to-medium term: a return to the use of carbon-dirty coal to fill the growing gap between energy demand and renewable-generated energy output. Perversely, Germany's original and well-intentioned pursuit of environmental idealism in the Climate Change cause has led it to the point where it is burning increasing quantities of carbon-dirty coal. It faces further stark choices: keep the Green dream alive and sacrifice global economic competitiveness as a consequence of high and ever-rising energy prices, or keep the lights and heating on, and prices down, but relinquish the idealistic vision of a renewable-energy fuelled future.

Focusing the attentions of Merkel's government and opposition parties are companies like the German chemicals conglomerate BASF, who between 2009 and 2013 shifted $5.7 billion of investments from Germany to the US.[46] The link between environmental aspiration and economic success was not lost on Philipp Rösler the Economy Minister, who recognised the damage caused to German industry by high energy prices: 'German companies that are deciding in favour of other locations and do not want to set up their business in Germany...The challenge is to promote and expand renewable energies without jeopardizing competitiveness.'[47] A definite shift in political priority from the environment towards individual benefit and economic competitiveness can be detected, despite the Green rhetoric. German attempts to ameliorate global warming through the pursuit of ambitious renewable energy projects and the associated reduction in carbon output have had a major unintended effect. By rejecting nuclear-generated energy after the Fukushima disaster, and opposing the development of shale gas and oil exploration, the increasing use of coal has resulted in a growing carbon footprint. In contrast, America's growing reliance on gas produced from shale rock – which burns more cleanly than coal – has led to a falling carbon footprint.

One consequence identified in the messages from key members of the Merkel government is a repositioning of environmental priorities. As with Cameron and the UK, harsh economic realities are pointing towards a political shift from idealistic environmentalism to a more economically-motivated pragmatism. The language of climate change and renewable energy is still present but it is falling behind economic priorities, much to the chagrin of the lobbyists who had fought so hard to advance the Climate Change cause and gain what looks increasingly likely to be only a temporary grip on policy making. Furthermore, the European Commission announced in January 2014 that it had rejected a legally binding target for renewable energy in Europe by 2030.[48] Cameron reinforced the downgrading of environmentalism in relation to the economy when he announced two days later: 'A key part of our long-term economic plan to secure Britain's future is to back businesses with better infrastructure. That's why we're going all out for shale. It will mean more jobs and opportunities for people, and economic security for our country.'[49]

Returning to President Obama, shortly after winning the 2012 Presidential election he gave a press conference that was much more illuminating than his highly scripted State of the Union speech a few weeks later. While still revealing a desire to uphold Climate Change ideals, it is possible to detect a shift in his priorities toward human interests from environmental concern. He was asked: 'What specifically do you plan to do in a second term to tackle the issue of climate change?' Obama answered: 'What we do know is the temperature around the globe is increasing faster than was predicted even 10 years ago.'[50] Critics pointed out the scientific inaccuracy of this claim: even Rajendra Pachauri, Chair of the IPCC, acknowledged in February 2013 that there has been a hiatus in global mean temperature rises since 1997.[51] Setting the temperature claim aside, however, Obama continued: 'I am a firm believer that climate change is real, that it is impacted by human behavior and carbon emissions.'[52] He then told the world where climate change stood in his list of political priorities:

> There's no doubt that for us to take on climate change in a serious way would involve making some tough political choices, and you know, understandably, I think the American people right now have been so focused and will continue to be focused on our economy and jobs and growth that, you know, if the message is somehow we're going to ignore jobs and growth simply to address climate change, I don't think anybody's going to go for that.[53]

Some may consider it unfair to attach much weight to an unscripted press conference, though it can also be argued that under pressure most people default to their most deeply held views. However, revisiting Obama's 2013 State of the Union Address, he makes his main concern explicit:

> Our first priority is making America a magnet for new jobs and manufacturing. After shedding jobs for more than 10 years, our manufacturers have added about 500,000 jobs over the past three. Caterpillar is bringing jobs back from Japan. Ford is bringing jobs back from Mexico. And this year, Apple will start making Macs in America again.[54]

During the course of his first term in office Obama's commitment to Climate Change waned; it is clearly a very long way from Tony Blair's 2006 statement to the UN that 'there is no issue that is more important than climate change.' For Obama and other leaders of recession-hit countries, Climate Change has fallen behind the 'economy and jobs and growth' as a political priority. The importance of projected future threats to the environment contained in the regime of climate truth is eroding in relation to the severity and imminence of economic threats. The extent of the rejection of environmental idealism in the political realm was confirmed at a global level when the Kyoto Protocol expired on 31 December 2012 with no agreed international commitment to replace it. For the staunch advocate, Climate Change has become an unquestioned and unquestionable article of faith whose moral hierarchy is shaped by post-normal science and its associated environmental idealism, and whose political priorities are non-negotiable: the natural world must be protected against the human enemy. Meanwhile, it is becoming increasingly obvious that for political leaders the value of Climate Change truth reflects the value of stocks, shares and global mean temperatures: it can go up as well as down.

Summary

The individual who seeks to act ethically in respect of human advancement, Climate Change and protection of the environment is left in an uncertain position: support the Green 'ideal' – prioritising care of the environment with its attendant global consequences for economic growth and personal wealth; ignore the environment and disregard Climate Change in pursuit of short-term human interests; or opt for a more pragmatic ethic that settles on a compromise somewhere between the maintenance of living standards and the well-being of the planet. If politicians drove forward climate initiatives in response to claims of an inevitable global warming crisis – aided and abetted by a loosening of normal scientific standards and an accompanying reliance on consensus-based, risk-centric, post-normal scientific assessments – the reverse can also happen and there are indications that it is already taking place.

In Europe the global warming-inspired Green agenda emerged at a time of rapid economic growth as a financial bubble expanded to give the impression of abundant wealth creation. A regime of climate truth swept all before it between the 1980s and the 2000s as Climate Change emerged as a global environmental cause. In that context the drive for sustainability – against the background of clear late-twentieth century global warming and siren warnings from the IPCC about the consequences and inevitability of relentless global temperature rises – appealed to politicians of almost every hue. A portion of economic growth could be affordably sacrificed by the rich for the greater good of the planet, while encouraging a personal feel-good factor in individuals who viewed themselves and their actions in ethical terms.

However, what politicians like Obama, Cameron and Merkel have begun to expose is the appealing myth that human advancement and environmental concern can both be attained in absolute terms. They cannot, and harsh choices must be made. President Bill Clinton's 1992 election-race mantra – [It's] 'the economy, stupid' – sums up the brutal reality of national and global politics in tough times: a deeply wounding truth to wield in the climate truth wars. If action to prevent global warming and climate change costs money and/or jobs, it is currently deemed too expensive. In the politics of climate truth, as elsewhere in life, actions speak louder than words. President Obama may talk good Climate Change and environmental concern but the noise in the background as he does so is from the never-ending convoy of trucks and equipment heading to and from the Dakota shale-fields. In the corridors of power, the economy has trumped Climate Change in the truth wars.

There is another, even more uncomfortable way to understand the subjective reality of the climate truth wars. The desire to combat global warming through economically restrictive activities could be seen as a rich person's (and a rich country's) self-indulgent luxury: prompted, ironically, by Climate Change's anti-capitalist ideology and motivated by repeated threats of impending climate-induced Armageddon. With the global economy currently facing multiple threats and global warming at an apparent standstill that commenced around 1997, *realpolitik* is placing economic concern and individual prosperity over climate policy and care of the environment. In the climate truth wars, political responses to severe short-term crises have marginalised any collective response to the long-term crisis promised – threatened even – by dire climate projections. Perversely, the relative 'success' of the Climate Change cause in European policy-making has led to a recent increase in CO_2 emissions, while its relative 'failure' (failure to stop environmentally harmful fracking) in the US has resulted in falling CO_2 levels.

Climate Change is under pressure from multiple competing truth claims: internal contradictions and challenges between normal science and ideologically-driven post-normal science; contested priorities over care for the environment and concern for human advancement; and external factors such as alternative global threats and competing political interests. National leaders, especially in the most highly developed and developing countries, have downgraded the amelioration of global warming and climate change in their priorities. The IPCC and environmental lobby groups continue their work but with less and less access to the corridors of power. If truth is produced in relations of power then objective climate truth – or at least its avatar – has gone for now. Climate Change advocates may have effectively constituted a regime of climate truth, however, it has been relegated behind more urgent 'truths' about the need for economic growth and a desire to maintain standards of living.

Part II
Politics, Truth and Military Intervention

4
Tyranny, Freedom, Democracy

'That guy is a tyrant!'

These words were uttered, not with disgust and rage but in the kind of hushed, reverential tones normally reserved for news anchors announcing the death of a global icon or forewarning the public about the imminent landfall of a hurricane of extreme magnitude. Instead, it was half time in a high school football game and the team coach was so incensed by the poor attitude and effort of two supposedly leading players that he sent them to run up and down the empty playing surface while he tried to rouse the rest of his losing team. The admiring 'tyrant' comment came not from a neutral observer but from the father of one of the boys running up and down the football field who had similarly been embarrassed by his son's poor first-half showing. He totally supported the public humiliation of his son by the coach whose hatred of losing was only matched by his apparent hatred of teenagers. Only in countries free of genuine tyranny could such a warped or misplaced understanding of the word be allowed to flourish. For example, Bill Shankly, one of the greatest football[1] managers of all time is attributed with the words: 'Some people believe football is a matter of life and death, I am very disappointed with that attitude. I can assure you it is much, much more important than that.'[2]

Contrast the *faux* tyranny of the sports field with the markers of true tyranny seen in recent decades: massacres from Rwanda to Srebrenica; ethnic cleansing in Kosovo; child soldiers in the Congo; rape as a weapon of systematic violence and oppression from Cambodia in the 1970s to Bosnia in the 1990s and on to the Democratic Republic of Congo in the second decade of the twenty-first century. From a liberal Western perspective, a crucial aspect of all major intervention discourses of the

77

past two decades is the desire, the *need*, for tyranny to be confronted and for tyrants to be removed if they are unwilling to change their ways. All of which presupposes a particular understanding of tyranny where it is not only cruel, destructive and barbaric but anti-democratic, anti-freedom and generally opposed to the central tenets of liberal ideology.

Shortly after the fall of the Iron Curtain, would-be President Bill Clinton described the kind of country he wanted to lead: 'An America that will not coddle tyrants, from Baghdad to Beijing. An America that champions the cause of freedom and democracy from Eastern Europe to Southern Africa – and in our own hemispheres, in Haiti and Cuba.'[3] Despite his noble words, however, like many other international leaders he was not prepared for the speed and ferocity of the 1994 Rwanda massacre and could only watch in horror as events unfolded. Then in 1997, President Clinton acquired a deeply committed, pro-intervention ally when Tony Blair was elected British Prime Minister. Reflecting on his approach to tyrants and their horrific acts when he came to office, Blair's unwavering view was very simple: if he assessed that a tyrant was inflicting unjust violence and untold suffering on his people then he [Blair], and the world, had a moral obligation to intervene to remove the regime concerned.[4]

The merits of democracy are, to many who live in democracies of various hues, indisputable: a truth beyond question based on principles beyond price. Which is difficult for the Westerner to conceive is that there are people and places for whom such a truth claim is an abomination. In secular, liberal democracies where freedom is seen as the crowning achievement of the individual, that freedom cannot be assumed to mean the same thing as 'freedom' in, say, a Muslim theocracy like Iran, or even in other less strict but nonetheless Islamic countries. For Muslims, personal belief systems cannot be separated into religious and secular, public and private, personal and political, in the way that it can for most Christians in liberal states. Islam pervades all areas of private and public life in ways that cannot be easily understood by Western observers.

In light of these observations, the remainder of this chapter will examine the relationship between military intervention, opposing tyrants and promoting freedom and democracy. Beginning by highlighting how opposing tyranny was used as part of the justification of military interventions by external third parties in Kosovo, Afghanistan and Iraq, discussion will move on to consider the extent to which tyranny has been replaced with freedom and democracy. Then finally, the situation in Syria and Libya between 2011 and 2014 will be analysed. By the time the use of chemical weapons caused international outrage in 2013, repeated military interventions had brought the US, UK and

NATO to the point of political, financial and humanitarian exhaustion. This exhaustion, in turn, prompts reflection on the universal applicability of democracy as a political system, and military intervention as a means of achieving it. The chapter will conclude with a discussion of the extent to which the pursuit of freedom prompted by liberal Western ideals inevitably founders on the democratic principles that underpin interventionism in the first place.

Opposing tyranny

In the West, opposition to tyranny is taken, unquestioningly, not only to be a moral good but also a moral obligation. Underpinning this attitude is the ideological claim – explicitly acknowledged in the US Constitution and taken for granted in the UK's unwritten constitution – that the pursuit of freedom and the exercise of democracy are not only the highest form of political system but also the personal ethic upon which civilised society depends. Winston Churchill expressed the sentiment in a more convoluted way when he told Parliament in 1947: 'Democracy is the worst form of Government except all those other forms that have been tried from time to time.'[5] Furthermore, the extension of freedom and democracy to those who suffer under tyrannical leaders in other forms of government is the ultimate *aim* of those who value freedom and the principles of liberal democracy.[6] While this chapter will progress under such an assumption, the reader should be aware that it is just that: an assumption, which can and should be contested. Such a benign, unquestioned view of tyranny overlooks the possibility that the advancement of liberalism itself can assume a tyrannical dimension if it sees every other opponent, political system or ideology as illegitimate and worthy only of subjugation, defeat and replacement.[7] John Adams, the second President of the United States, cautioned two centuries ago:

The right of a nation to kill a tyrant, in cases of necessity, can no more be doubted, than that to hang a robber, or kill a flea. But killing one tyrant only makes way for a worse, unless the people have sense, spirit, and honesty enough to establish and support a constitution guarded at all points against tyranny; against the tyranny of the one, the few, and the many.[8]

Tyranny is a word easily projected onto an enemy or rival who looks, acts and thinks differently to me – based on cultural, religious and other values I do not recognise or understand – sometimes with full justification and occasionally with less obvious cause. Tyranny is not the opposite of liberalism, a mere enemy to be overcome; it is also the temptation liberalism must guard against as it advances its cause. And one such cause was pursued by NATO as the Serb leader President Slobodan Milosevic embarked upon a campaign of ethnic cleansing against Kosovar Albanians in 1998 and 1999.

By 24 March 1999 NATO aircraft had been conducting an aerial bombardment of the Federal Republic of Yugoslavia for four weeks in an attempt to coerce Milosevic into stopping the atrocities that his regime was inflicting upon the people of Kosovo, a region of Greater Serbia that had previously enjoyed significant autonomy for many years. President Bill Clinton addressed the American people to explain why American aircraft were involved in the NATO action. He began by highlighting the record of warmongering, oppression and violence on the part of Slobodan Milosevic, the leader who had deployed Serb forces against Bosnia, Croatia, Slovenia and now Kosovo in the preceding decade.[9] Having rejected the offer of a peaceful solution to the Kosovo dispute, Milosevic's forces stepped up their offensive against Kosovar Albanians. Clinton stated:

> Now they've started moving from village to village, shelling civilians and torching their houses. We've seen innocent people taken from their homes, forced to kneel in the dirt, and sprayed with bullets; Kosovar men dragged from their families, fathers and sons together, lined up and shot in cold blood. This is not war in the traditional sense. It is an attack by tanks and artillery on a largely defenceless people whose leaders already have agreed to peace. Ending this tragedy is a moral imperative.[10]

A month later on 24 April 1999, NATO heads of state had gathered to celebrate 50 years since the formation of the alliance to assess the effectiveness of the ongoing strategy in Kosovo and to consider what the future might hold. Against that backdrop, Prime Minister Tony Blair made one of the boldest political speeches of modern times, a speech that went much further than President Clinton's and which would help shape Western military interventionism in subsequent years. It was only a few years after the end of the Cold War, during a decade where Francis Fukuyama's 'end of history'[11] thesis (which has been strongly contested ever since) appeared to confirm Western liberal democracy as the inevitable high point of political evolution, that Blair proposed an increasingly assertive liberalism. Just as economic liberalism was driving globalisation and wealth creation, Blair argued that powerful liberal democratic states – to be led by the US and supported by the UK – should be prepared to use military force to advance the cause of freedom[12] against totalitarian, oppressive regimes and leaders. In that context, he described some of the atrocities being committed by Serb police, militia and soldiers at the command of Slobodan Milosevic: 'Awful crimes that we never thought we would see again have reappeared – ethnic cleansing, systematic rape, mass murder,' which had resulted in 'the tear stained faces of the hundreds of thousands of refugees streaming across the border.'[13]

The complexity of the intertwined histories of Serbia and Kosovo does not lend itself to pithy summary or media sound bite. Much more effective in prompting and maintaining support for military action to prevent these abominations against the innocent is the use of the 'baddie'. The subtleties of claim and counter-claim in distant lands going back in time for 800 years or more tends not to tug at the heartstrings or drum up political support for war – and a willingness to pay the costs of war – either in the American southern bible-belt or the industrial rust belt in the north, in the British midlands or among London's metropolitan elite.

In contrast, the 'baddie' has been constituted as 'evil' in numerous Western cultures for centuries, spanning the years from Shakespeare to Hollywood. Superficial confrontations between good and evil have played out countless times in the cinematic format: from Clint Eastwood's fictional 'Man With No Name' to the everyday heroes of United Airlines Flight 93 who died while tackling Al-Qaeda terrorists on that fateful day in September 2001. For Blair, confronting an evil foe needs little or no justification: it goes beyond mere social expectation to the extent that he sees it as an unquestioned, inherent good. His evil enemy merely needs to be identified and action taken by good men and women, their 'goodness' being conferred by the very action of confronting the evil one – literally or figuratively. However, such a simplistic approach overlooks the possibility that what looks like evil, and indeed how it should be addressed, from a liberal Western perspective may be viewed differently from within other cultures or political systems. On 9/11, news channels showed citizens in several parts of the Middle East celebrating the destruction of the Twin Towers and the humiliation of the United States. Denouncing those individuals as 'evil' unhelpfully avoids the need for cross-cultural understanding when it classes the celebrators as less than human. Furthermore, such a one-dimensional approach to good, evil and tyranny also does not explain how some leaders that Blair and many others would class as tyrants often manage to enjoy strong – even if not universal – public support in their own countries.

Blair used the good-evil dichotomy in a simplistic way as he sought popular and political support for a new doctrine of humanitarian intervention in response to two specific tyrannical rulers:

> We cannot let the evil of ethnic cleansing stand. We must not rest until it is reversed. We have learned twice before in this century that appeasement does not work. If we let an evil dictator range unchallenged, we will have to spill infinitely more blood and treasure to stop him later ... Many of our problems have been caused by two dangerous and ruthless men – Saddam Hussein and Slobodan Milosevic. Both have been prepared to wage vicious campaigns against sections of their own community ... If NATO fails in Kosovo, the next dictator to be threatened with military force may well not believe our resolve to carry the threat through ... War is an imperfect instrument for righting humanitarian distress; but armed force is sometimes the only means of dealing with dictators.[14]

Blair was, of course, highly selective about the dictators he was willing to use force against. President Robert Mugabe was studiously ignored despite many years of oppressing segments of the Zimbabwean population, a combination of post-colonial guilt and a lack of regional allies putting intervention beyond political possibility. One man who could not be ignored, however, wrote his place in infamy when his Al-Qaeda acolytes destroyed the Twin Towers in 2001; American immunity from religious and ideological struggles half a world away crashing down with the steel and concrete in New York.

Shortly after the 9/11 attacks President Bush had reinforced the popular 'evil baddie' representation of bin Laden when he invoked the response of the classic Hollywood Western: 'I want justice. And there's an old poster out west, that I recall, that said, "Wanted, Dead or Alive".'[15] A few days later in his address to a Joint Session of Congress on 20 September 2001, President Bush described the actions of Al-Qaeda terrorists carried out at the command of Osama bin Laden and motivated by his political-religious ideology:

> The terrorists' directive commands them to kill Christians and Jews, to kill all Americans and make no distinctions among military and civilians, including women and children ... There are thousands of these terrorists in more than 60 countries. They are recruited from their own nations and neighborhoods and brought to camps in places like Afghanistan where they are trained in the tactics of terror. They

are sent back to their homes or sent to hide in countries around the world to plot evil and destruction.[16]

'Evil', anti-Christian, anti-Jew, indiscriminate killers: these are bin Laden's terrorists in Bush's rhetoric. Yet – and shocking to Western sensibilities – elsewhere these were well-educated, idealistic heroes and martyrs. Their very descriptions marked a crucial front in the truth wars at that time.

In the weeks and months that followed – which included a rapid toppling of the Taliban regime in Afghanistan – bin Laden began to disappear from President Bush's rhetoric, to the extent that he was challenged on the absence by a journalist in March 2002. Bush responded: 'I don't know where [bin Laden] is...I truly am not that concerned about him. I know he is on the run. I was concerned about him, when he had taken over a country.'[17] The problem with representing Osama bin Laden as an amalgam of an evil leader, demi-religious guru and murderous, terrorist figurehead is that the failure to capture or kill him would become an increasing source of embarrassment to the world's most powerful individual and the institutions he commanded. It is impossible to tell how the history of the first decade of the twenty-first century might have been written had the Taliban handed Osama bin Laden and his followers over to the American authorities. Even the drone wars may have taken a different turn if the Pakistan government and security services had located, arrested and surrendered the fugitive. Instead, the biggest, costliest, most violent manhunt in history lasted almost ten years before bin Laden was shot dead by a US Navy SEAL at home in his Abbottabad compound. In the meantime, President Bush responded in two ways: first by increasingly emphasising the wider Al-Qaeda terror threat rather than focus on its elusive leader; and second, by offering up an even more notorious baddie – President Saddam Hussein. As he shifted his focus from a man who had done significant direct damage to the US, Osama bin Laden, to one who had not, Saddam Hussein, Bush found a staunch and articulate ally in Tony Blair.

The tenor of Blair's reference to Saddam Hussein in his 1999 Chicago speech captured the revulsion and distrust in which the British leader held his Iraqi counterpart. However, it served only as a foretaste of the way in which Blair would describe Saddam's character as evil and vindictive in his speeches in the years that followed. As the Bush administration turned its gaze from the mountains of Afghanistan, Al-Qaeda and the deposed Taliban regime towards Saddam Hussein and Iraq, Blair – having committed himself and his country to standing shoulder-to-shoulder with the US[18] – wasted no time in devoting himself to the American cause. The Prime Minister's support was appreciated by President Bush, who continued with the good *versus* evil theme when referring to his ally: 'I've got no better person I would like to talk to about our mutual concerns than Tony Blair. He brings a lot of wisdom and judgment, as we fight evil'.[19] In 2002 the two leaders met at Camp David to agree a way forward in dealing with Saddam Hussein. As he prepared for the summit that would shape the politics of the two countries for the next decade Blair made his feelings known to his close advisors in London. Despite everything Blair said publicly about ridding the world of Iraq's weapons of mass destruction (WMD), his motivations were more focused on the dictator he wanted to remove. On 17 March 2002 he wrote to his Chief of Staff:

> Saddam's regime is a brutal, oppressive military dictatorship. He kills his opponents, has wrecked his country's economy and is a source of instability and danger in the region … a political philosophy that does care about other nations – e.g., Kosovo, Afghanistan, Sierra Leone – and is prepared to change regimes on the merits, should be gung-ho on Saddam.[20]

Unfortunately for Blair, when he stood alongside President Bush at the Camp David press conference almost three weeks later, the niceties of international law as set out in the UN Charter, coupled with legal guidance from his Attorney General, stopped him from calling for regime change in Iraq or from sharing in public the strength of his views about removing Saddam that he voiced in private. In contrast, President Bush made his position clear in a way that put Blair in a difficult position:

> This guy, Saddam Hussein, is a leader who gasses his own people, goes after people in his own neighbourhood with chemical weapons, he is a man who obviously has something to hide … I explained to the Prime Minister that the policy of my government is the removal of Saddam.[21]

Sensing a gap between the American and British positions, a journalist asked Blair 'if it is now your policy to target Saddam Hussein?' in the process also reminding the Prime Minister that Heads of State are normally considered immune from personal attack.[22] In the reply that followed, Blair demonstrated the malleability of truth in the hands of a master of political rhetoric. Despite his privately-aired view that he was prepared to change regimes 'on their merits' and that his political philosophy prompted him to be 'gung-ho' on Saddam, Blair equivocated, telling the world: 'You know it has always been our policy that Iraq would be a better place without Saddam Hussein.'[23] This may have been Blair's opinion, it may have been a political aspiration, or it might even have been wishful thinking. However, it was not a policy. Blair was either deceiving himself or deceiving the public in the process of trying to support the American position. President Bush even recognised that he had put his ally in the battle against evil and tyranny in a difficult position, adding: 'Maybe I should be a little less direct and be a little more nuanced and say we support regime change.'[24]

In the year that followed, political wrangling escalated, UN weapons inspectors were finally given access to alleged sites of WMD production in Iraq, and the Bush administration maintained its stated aim of removing Saddam. Meanwhile, Blair had to negotiate a number of contradictory positions: supporting an ally with whom he agreed but who was not constrained in the ways he was; leading a political party that did not support his view; promoting democracy while ignoring the two million people who marched against the war in London and the half of the country that opposed war; making a case for war while accommodating an Attorney General who consistently opposed the Iraq intervention until the eleventh hour. Opposing tyranny was the only part of Blair's case for war that was not consistently undermined by legal opinion or the inconvenient lack of evidence of WMD.

By mid-March 2003 all the chess pieces were in place: American forces were ready to strike against Iraq and British forces were in place to support them. Blair eventually made his personal view public as war drew near and military withdrawal became politically and logistically near impossible. On 16 March 2003 he announced: 'The Iraqi people deserve to be lifted from tyranny and allowed to determine the future of their country for themselves.'[25] On the eve of the invasion, Blair sought support for war from the British Parliament in a speech that was almost entirely focused on Saddam's personal malevolence, which remained beyond dispute even if WMD were proving elusive. Saddam was a man who 'brutalised' others, using WMD against Iran and his own people. He deceived Kofi Annan and the UN's weapons inspectors through 'lies, deception and obstruction' and 'playing the same old games'.[26] Saddam's tyranny had led to 60 per cent of its population needing Food Aid; thousands of children dying from hunger and lack of basic medicines; four million Iraqis in exile; death and torture camps; mutilation for enemies of the state; and so on.[27]

It took only a few weeks until Saddam's regime was toppled and he fled for his life. The 'Ace of Spades' – as Saddam was referred to in the playing-card list of Iraq's 'most wanted' – was eventually found hiding in a fetid underground cell. Granted a trial that was denied to his many victims, he was tried and subsequently hanged. There is no need to recount here the many political miscalculations that led to the US-led coalition snatching chaos from the jaws of victory as Iraq descended into an inter-tribal, sectarian insurgency. Suffice it to say that the credits did not roll as they do on a cinema screen, evil was not curtailed by Saddam's death and, despite President Bush's victory declaration on 1 May 2003, in the Iraqi tragi-drama it was not 'The End'.

After US and UK forces eventually withdrew from Iraq after several years of costly counter-insurgency fighting, the death of Osama bin Laden arrived as an almost unexpected postscript in the Afghanistan War: testament to the dedication and resources of the American Intelligence and other services that eventually tracked him down to his bolt-hole in Pakistan. By May 2011 Barack Obama had replaced George Bush as President but the fight against terror(ism) continued unabated. Obama had already authorised hundreds of drone strikes against Pakistani Taliban and Al-Qaeda targets – many more than Bush had – but this was different. Previous violations of Pakistan's airspace were by remotely piloted aircraft: the bin Laden operation would involve a much more significant incursion into an ally's sovereign territory. The insertion of a US Navy SEAL team in two Black Hawk helicopters would be supported by additional combat troops nearby, search and rescue helicopters if needed, surveillance pictures from drones, and the full range of Air Force strike aircraft capability.[28] The Pakistan government was not consulted in advance in case an intelligence leak allowed bin Laden to escape again. In the event, despite the inevitable complications when the fog of war descends, the SEAL team was successful. Subsequent increased tensions between the US and Pakistan were a small price to pay for the political opportunity the death of bin Laden afforded President Obama. The baddie was dead, the sheriff had brought his man to justice, and he could now move more decisively towards ending America's involvement in an increasingly unpopular war.

For freedom and democracy

At the peak of his acting powers in 1995, Mel Gibson was chosen to play the thirteenth century Scottish leader William Wallace as he fought against English dominance in the film *Braveheart*. The film became best known for three things: Gibson's travesty of a Scottish accent; a mangling of historical events that was poor even by Hollywood standards; and Wallace's final line as he inspired a rag-tag army of ill-equipped Scots to fight, and probably die, against the overwhelming numbers of the English army. From all around the world there came reports of cinema audiences cheering and applauding – probably diaspora Scots like me – Wallace's stand against tyranny and his cry for 'FREEDOM!' People who knew nothing of Scotland and its history recognised the hunger for freedom: freedom from oppression and control – what Isaiah Berlin called 'negative liberty'[29] – and a willingness to make great sacrifices to achieve it.

Those who cherish freedom the most are usually those who paid the highest price or endured the longest suffering to achieve it, and *Braveheart* tapped into what for many people is a deeply-held longing. What is frequently overlooked, however, in the desire to impart freedom and democracy to those who currently live under the dominion of tyranny is the blood price that was paid for those freedoms over centuries by the countries and people who now enjoy them. Briefly consider the histories of the UK and US.[30] Wars such as those fought by William Wallace between the Scots and the English occurred over more than 500 years before Scotland and England even assumed anything like their modern geographical boundaries and the 1707 Acts of Union brought them together to form the Kingdom of Great Britain. And these regular blood-lettings did not even include the clan rivalries in Scotland or the domestic battles in England.

Later in the eighteenth century across the Atlantic Ocean, American colonists fought a bloody War of Independence that would lead to the birth of the United States of America. Though only around 50,000 American military men died in that war, as a proportion of the population at that time, the losses were substantial. However, the path to a contemporary understanding of freedom and democracy did not finish with the Declaration of Independence in 1776, it only started. The American Civil War almost a century later would cost more than 500,000 lives – more than the US would lose in either of the two twentieth century World Wars. Crucially, it brought a peace and political stability that remains today and which eventually led to full equality between citizens and voting rights for all Americans in the twentieth century. Those who would try to 'give' freedom and/or democracy to oppressed people in other countries naïvely overlook how long it took – and the awesome price paid – to achieve what they now take for granted. Further, the ideological superiority of Western notions of freedom and democracy is now so deeply inured in those who have enjoyed it for generations that it is the unquestioned political truth on which all others are based. It therefore becomes extremely difficult in the truth wars surrounding military intervention to conceive of legitimate alternative possibilities. The oratory of leaders like Clinton, Blair, Bush and Obama has encouraged Americans and Britons to equate ethical conduct and attitudes with supporting efforts to extend, by military force, the freedom and democracy they enjoy to others: Iraqis, Afghans, Libyans, Syrians and so on. Which begs the paradoxical question: if freedom is given to or enforced upon another, to what extent can it truly be called freedom? What about *their* freedom to choose a different polity, *their* freedom not to choose liberal Western democracy?

Returning to *Braveheart*, that two-hour, escapist, cinematic pursuit of freedom is more problematic than helpful when trying to assess the extent to which opposing tyranny – in the guise of Slobodan Milosevic, Osama bin Laden and Saddam Hussein – has resulted in the freedom and democracy espoused by Clinton, Bush, Blair, Cameron and Obama. The first obvious difficulty is overcoming the modern liberal assumption that democracy and freedom are synonymous. Freedom from oppression (negative freedom) and freedom to vote (positive freedom) are not one and the same thing. William Wallace may have fought for freedom against the English King Edward in the 1290s but he was not attempting to win universal suffrage for the common man. Wallace was a minor nobleman, one of over a thousand who owned the land on which peasant farmers eked out a harsh existence. He and his ilk were happy for commoners to be dominated as long as it was not English noblemen or kings who did the dominating. At least, however, those distant Scots were fighting for their own freedom, of a sort.

Consider the extent to which ending tyranny has resulted in freedom and democracy for Kosovars, Afghans and Iraqis. On 11 March 2006, former Yugoslav President Slobodan Milosevic died in custody while being tried for war crimes at the International Criminal Tribunal for the former Yugoslavia (ICTY) in The Hague. Seven years later in April 2013 Kosovo and Serbia finally signed an agreement to normalise relations between the two countries, with two caveats. First, Serbia still does not recognise Kosovo as a sovereign state; and second, Kosovo does not have a unified population, with the Serbs in its north being granted a measure of autonomy within the country: a mirroring of the position once occupied by Kosovars within Greater Serbia.[31] Kosovar-Albanian freedom has resulted in a limit on the freedom of their Kosovar-Serb counterparts. Despite this, the move allows both countries to advance their respective ambitions to become members of the European Union and by 2014, 15 years after the Kosovo War; around 100 UN member states had recognised the Republic of Kosovo.[32] Although there are continuing tensions between Serbia and Kosovo, unresolved claims and counter-claims regarding sovereignty and ongoing war crimes trials, a degree of political freedom and stable democratic structures have been achieved in the new Republic.

In contrast, the political situations in both Iraq and Afghanistan in 2014 are an unstable mix of social, economic and security failures, with the occasional success that offers hope for the future. In the north of Iraq, Iraqi Kurds are enjoying high levels of inward investment, expanding oil exploitation and relative peace and prosperity compared to their southern counterparts, presided over by the Kurdistan Regional Government (KRG) in the autonomous region. Falah Mustafa, KRG Minister for Foreign Relations, asked in 2013: 'Are we for an Iraq based on the constitution for genuine partnership and power sharing? Or are we for a totalitarian, authoritarian, one-man rule of Iraq again? As far as we are concerned, no (to the latter)! Definitely no!'[33] In relation to the deposed dictator, Yasin Mustafa Rasul explicitly connected the end of tyranny with the birth of freedom: 'Saddam is like a part of distant history or a fading bad dream ... We've almost forgotten that era. Freedom may be more important than wealth, but we have both.'[34] However, that 'freedom' can only be partial because the Iraqi Kurds remain part of Iraq and not part of a free and independent Kurdistan.

The success that has been achieved in the Kurdish region of Iraq has not been emulated elsewhere. There was relative calm in Iraq in late 2012 when fewer than 300 Iraqis were dying per month from sectarian violence, the lowest level since the outbreak of war in 2003. A year later, more than 1,000 Iraqis were dying each month in an escalating series of attacks: a level not seen since 2008 at the height of the insurgency.[35] As 2013 gave way to 2014 Iraq was again under siege as Sunni militants linked to the former Al-Qaeda affiliate Islamic State of Iraq and the Levant (ISIS) fought government forces for control of the country. Complicating matters, an unstable Iraqi government is presiding over regional rivalries, escalating Sunni-Shia tensions and an uneven distribution of oil wealth. Whatever abuses Saddam Hussein carried out – and he carried out many – he did not cause these rivalries when he ruled over Iraq; he exacerbated some tensions but supressed others in the way, for example, he abused the southern Shia and the Marsh Arabs after the 1991 Gulf War. Through brute force he did not allow these historically ingrained divisions to threaten his own regime. It is still possible in the future that the Iraqi people will sustain some form of freedom, including a democratic element, that some, perhaps many, of its citizens want. However, 'freedom' in the Islamic State covering swathes of Iraq and Syria and which was declared by ISIS on 30 June 2014 – freedom to impose a form of Islam that will 'liberate' Iraqis from impure Western influences – will look very different to Iraq's fragile, nascent 'democracy'. In Afghanistan, 2014 is a year of transition as US and UK forces prepare to withdraw back to barracks in their own countries. Or rather, US and UK military personnel are being withdrawn from a politically unpopular war so that both governments can push ahead with military cutbacks as they each try to control their spiralling national debts. Opposition to tyranny, like the pursuit of freedom and democracy for other countries who do not have it, comes at a heavy cost – in financial as well as human terms. In the truth wars, the *realpolitik* of saving the British and American economies – and by implication keeping their respective leaders and governments in office – has for the foreseeable future triumphed over the ideological pursuit of the spread of liberal democracy and the ethical arguments put forward as the interventionist wars started.

Consider some of the achievements of the vast outlay by the US-led NATO operation. According to the UN, 2013 was 'a record-setting year of poppy production and cultivation' in Afghanistan that resulted in 5,500 tonnes of opium that would have health and security implications in the country and beyond.[36] The UN also warned that Afghanistan ranked 175th out of 177 on the Transparency International Corruption Perceptions Index.[37] Combined, these outcomes highlight the hubris of Tony Blair when he said of Afghanistan in a speech at the George Bush Senior Presidential Library on 7 April 2002:

The Taliban are gone as a government. Al Qaida's network has been destroyed in Afghanistan, though without doubt a residual capability remains and we should still be immensely vigilant. The Afghan people feel liberated not oppressed and have at least a chance of a better future.[38]

Similarly, President Bush's comments about Iraq, aboard USS Abraham Lincoln on 1 May 2003, exhibits naïve idealism:

Today we have the greater power to free a nation by breaking a dangerous and aggressive regime...The transition from dictatorship to democracy will take time, but it is worth every effort. Our coalition will stay until our work is done and then we will leave and we will leave behind a free Iraq.[39]

Bush's hopes and aspirations would ultimately be corroded by painful experience of the limits to which free democrats are prepared and able to go to extend the same opportunities to Iraqis, Afghans and others. Even more problematic for the spread of freedom and democracy is the reluctance of many of those mentioned to embrace the political ideologies of what they see as conquesting forces.

There are a number of reasons for the differing outcomes in Kosovo, Iraq and Afghanistan beyond the success or otherwise of the military campaigns that have been waged. In April 1999 one of the conditions for military, or humanitarian, intervention proposed by Blair was: 'are we prepared for the long term? In the past we talked too much of exit strategies. But having made a commitment we cannot simply walk away once the fight is over.'[40] In the case of Kosovo, on 10 June 1999, as air operations against Serbian forces were concluding, the UN Security Council passed Resolution 1244 which set out the principles for a sustained peace.[41] These principles were largely respected by protagonists on both sides, Serb and Kosovar, as well as by the allies of the contending parties: Russia, in the case of the Serb authorities; NATO in the case of the Kosovars. Further, neighbouring states were committed to a sustainable peace that would provide stability in the region over the long term. The support of the United Nations, NATO, the European Union and regional neighbours over a 15-year period, combined with a desire by Serbs and Kosovars to build a better future for their now separate countries, has led to considerable optimism for the future. This, despite differences over the question of Kosovo's sovereignty that will probably never be fully resolved.

Afghanistan, over a 13-year period, has also been supported by the UN and NATO, with the EU donating around US$2 billion as well. More problematic, however, have been the competing political claims between multiple contending parties in the country, including warring tribal factions, corrupt central and regional governance structures, indigenous Taliban insurgents and trans-national Al-Qaeda terrorists. Exacerbating this volatile situation are wider regional factions, with huge amounts of weaponry making its way into Afghanistan via Iran on one border and via Pakistan on another. The historical rivalries, distinct cultures, diverse religious influences and competing political interests of the tribal groups that make up Afghanistan have mitigated against a 'successful' outcome – from a Western perspective – from the very beginning of the war in 2001. What currently passes for democracy in Afghanistan is one of the most corrupt political systems in the world, and prevailing notions of freedom are as likely to mean freedom from the US, NATO and their liberal Western ideas as freedom from the Taliban or other internal threat, complete with the religious restrictions that so constrained the population – especially women – in the past.

After 12 years of fighting for democracy and freedom in Afghanistan, President Karzai rewarded the sacrifices of the US, UK and allies by calling Americans 'rivals' and referring to the Taliban as 'brothers'.[42] Only a few days later, against the advice of the US, he released 65 Taliban prisoners who were accused of being responsible for the deaths of Americans and Afghans.[43] In weighing up the competing factors that shape President Karzai's political reality, he has concluded – probably sensibly given the circumstances – that since US and NATO forces will not be around much longer to protect him, his Administration and his successor, he needs to build bridges to those who pose the greatest threat to him. Anyone who seeks an alliance with the Taliban has already surrendered any pretence that they are pursuing any liberal Western conception of freedom.

Libya, Syria and beyond

After 20 years of post-Cold War military interventions by Western powers, half of which were prompted and intensified by the events of 9/11, 2011 was the year when US-led opposition to tyranny and support for freedom and democracy in distant lands reached a watershed. In the interventionist truth wars, political pragmatism began to openly usurp the idealism that had previously been voiced from Washington to London and Paris, and beyond. Uprisings against brutal dictatorships in both Libya and Syria presented America – supported inevitably by the UK but now also by France – with another opportunity to display its anti-tyrant, pro-freedom, pro-democracy credentials. Uprisings in Benghazi, Libya were followed very closely by protests in Daraa, Syria: both of which were preceded by the so-called Arab Spring that spread rapidly across several countries in North Africa and the Middle East. In keeping with their personalities and previous policies, Colonel Moammar Gaddafi[44] and President Bashar al-Assad responded aggressively to the uprisings that threatened the stability of their respective countries and the security of their regimes.

In response to events in Libya, President Obama once more included opposition to tyranny and the pursuit of freedom in his interventionist discourse:

For more than four decades, the Libyan people have been ruled by a tyrant – Muammar Qaddafi. He has denied his people freedom, exploited their wealth, murdered opponents at home and abroad, and terrorized innocent people around the world...Qaddafi chose to escalate his attacks, launching a military campaign against the Libyan people...we must stand alongside those who believe in the same core principles that have guided us through many storms: our opposition to violence directed at one's own people; our support for a set of universal rights, including the freedom for people to express themselves and choose their leaders.[45]

Giving an indication of the extent of America's willingness to intervene, the President was careful to stress that 'America's role would be limited, that we would not put ground troops into Libya'.[46] True to his word, after the initial air operations against Libya the US handed over operational command to NATO – led primarily by France and the UK in this instance – who took responsibility for protecting the Libyan people from their leader, Gaddafi. Uniquely in interventions since 1991, the UN Security Council authorised military action, with the proviso that only air power could be used to protect Libyan civilians.[47] French and British strike aircraft led the fight against Gaddafi's regime with the US providing specialist support: intelligence, logistics, electronic jamming and search and rescue.[48] Obama made explicit his reasons for America's limited involvement in intervention against a dictator who had killed almost 200 Americans in the 1988 Pan Am Flight 103 Lockerbie Bombing: to reduce risk and cost 'to our military and to American taxpayers'.[49] In this particular battle, idealism managed to conduct a few sorties against the Libyan regime before Capitol Hill pragmatism grounded the jets again.

Over the subsequent months, NATO effectively provided the air power component of the Libyan opposition – or rebel – ground forces, the regime was swept from power and Gaddafi was shot near a drainage culvert where he had been hiding. Two years later however, despite democratic elections in 2012, Human Rights Watch reported that governing structures were fragile, with armed militia making the achievement of law and order impossible.[50] Centralised tyranny has been replaced by randomised, localised tyranny at the hands of former revolutionaries who refuse to disarm.

In parallel with events in Libya, the uprising in Syria was gaining ground, prompting a severe military response from President Assad and his regime. By 2012 tens of thousands of Syrians had been killed in an escalating civil war – far more than had died in Libya – prompting President Obama to declare: 'We have been very clear to the Assad regime – but also to other players on the ground – that a red line for us is we start seeing a whole bunch of chemical weapons moving around or being utilized.'[51] By the time chemical weapons were used in August 2013 against Syrian civilians in Ghouta, over 100,000 Syrians had been killed and over two million refugees displaced either internally or externally.

So where was the intervention by the US, UK, NATO or anyone else? The severity of the oppression conducted by the Assad regime matched or exceeded that of any of the tyrants who had been deposed or killed in recent years: Slobodan Milosevic, Osama bin Laden, Colonel Gaddafi, Saddam Hussein. Obama's 'red line' threat about chemical weapons came back to haunt him as he and allied world leaders struggled to agree a coherent response to events in Syria. Several factors made the situation more complex than preceding interventions. First, Syria's main ally, Russia, made it clear that Assad would receive wholehearted Russian assistance in the event of an air or ground assault against it. Second, Syrian opposition groups proliferated at a rate that Western intelligence could not keep up with. Even when the National Coalition for Syrian Revolutionary and Opposition Forces was formed in November 2012 out of several opposing factions, it did not include the ever-expanding militant Islamist groups who seek an Islamic state. Third, it therefore became impossible to articulate a desired political end-state for Syria – from a Western perspective. Fourth, despite the atrocities reaching a level that would provide a just cause for intervention in any reasonable application of just war criteria, the other just war criteria of last resort, legitimate authorisation of force and reasonable possibility of success looked more elusive by the day. Fifth, and most importantly, the hearts of the American and British leaders and their people were not in it.

President Obama could stand in Sweden at the height of the controversy and say, 'we stand up for universal human rights, not only in America and in Europe, but beyond,' and sound as eloquent as ever.[52] However, his words had become mere words: the American fist that had reinforced similar sentiments a decade before had been unclenched. When the British parliament put military intervention in Syria to a vote on 30 August 2013, it was the first time in three centuries that MPs voted against a Prime Minister on the matter of going to war. That same week, a poll showed that only *eight* per cent of Britons polled supported immediate air strikes against Syria.[53] The next day President Obama faced a similar lack of support from the American public and asked Congress to approve military action, the immediate result of which was a delay in making any decision and the secondary effect of which was to share responsibility – and therefore blame – with Congress if war went ahead. In the event, an unforeseen series of political negotiations between the US, Russia and Syria led to the Assad regime agreeing to give up its chemical weapons. For all the talk of opposing tyranny and promoting freedom and democracy in the corridors of power in Washington and London for the past two decades, the Syrian civil war continues (at time of writing) to proceed with no indication of military intervention in pursuit of freedom by American, British, French or other NATO forces. In contrast, Islamist groups are increasingly gaining weapons, power and support in their pursuit of freedom *from* what they see as the corrupting influence of Western liberal democracy.

Summary

The notion of opposing tyranny may have provided an immediate boost in public support for the military interventions started by Clinton, Bush and Blair. However, hard experience has demonstrated how this partic-ular discourse runs a serious risk of overstating the extent to which major geo-political problems can be solved – and liberal Western ideology advanced – by the removal or death of one individual, even in the pursuit of freedom and democracy. At the Munich Security Conference on 1 February 2014, British Defence Secretary Philip Hammond, acknowl-edged what has been obvious for several years – that the public in most Western countries was 'war weary' after a decade and more of fighting in Afghanistan and Iraq.[54] Tried and trusted methods of inciting support for wars – such as stressing the need to confront terrorism abroad to make life safer at home – were being flatly rejected. Two claims stand out: first, his claim that 'There is a climate of scepticism about engage-ment in failed or failing countries, a fear of getting entrapped in longer term, deeper forms of engagement'; and second, that 'There is definitely a fear, quite irrational in some cases, that any engagement anywhere will somehow lead to an uncontrollable commitment to large numbers of troops, a large amount of resources and a long term intervention.'[55] The first of these claims appears wholly accurate. A CNN poll in December 2013 showed only 17 per cent of Americans supported the ongoing war in Afghanistan,[56] echoing similar views in Europe.[57]

More problematic is the subsequent statement that the public fear of military intervention is in any sense 'quite irrational'. It is entirely rational. The populations of NATO countries from the US to Europe have spent more than ten years watching coffins return home and billions being spent with little obvious and positive lasting effect. The

78-day aerial war over Kosovo and the liberal interventionist idealism of Clinton, Blair and their fellow NATO leaders seems a lifetime away. Even the trauma of 9/11 no longer sufficiently fuels American public outrage enough to want to keep fighting, especially since the tyrants at the heart of the Kosovo, Afghanistan, Iraq and Libyan interventions are long dead. Whatever ideological claims underpinned the initial rush to military intervention have been replaced by a more urgent truth: people in the West might fight for their own freedom but they now have little will to fight for the freedom of others. Especially since there is no evidence that any substantial freedoms have been gained from more than a decade of fighting. President Obama and Prime Minister David Cameron's willingness to conduct military intervention has been curtailed by the representative democracy that they and their predecessors sought to export: their people have spoken and said 'enough is enough'.

Worse, the standing of the US and UK in the places where the fighting has been the most severe is at its lowest ebb since the distant wars began. Far from voicing appreciation for the efforts of NATO personnel against Taliban and Al-Qaeda forces in February 2014, as ongoing NATO involvement in Afghanistan stumbled towards its conclusion, President Karzai was highly critical of the conduct of US troops in his country and disparaging of British efforts in Helmand.[58] In the intervention truth wars, liberal idealism – opposing tyranny while promoting freedom and democracy – has led soldiers into battle. However, it is the exhausted drumbeat of *realpolitik* that marches the surviving NATO-soldiers home again. Until the US and UK overcome their current war weariness, or a direct attack provokes retaliation against a new enemy, the people of Syria and anyone seeking help to escape tyranny or to further Western-style democracy in the near future will have to look elsewhere for their salvation.

5
Gendering Military Intervention

At the time of the Kosovo intervention in April 1999, Tony Blair, as British Prime Minister, was deeply committed to what he called humanitarian intervention. At a NATO summit in Chicago in that month he set out his personal concerns for the victims of oppressive regimes and proposed political responses in moral terms:

> We cannot turn our backs on conflicts and the violation of human rights within other countries if we want still to be secure...the principle of non-interference must be qualified in important respects. Acts of genocide can never be a purely internal matter. When oppression produces massive flows of refugees which unsettle neighbouring countries then they can properly be described as 'threats to international peace and security'.[1]

Despite Blair's idealistic rhetoric, wherever intervention was called for in the years that followed, competing national interests and the complexities of international politics precluded the possibility of any objective analysis of events on the ground, making collective, universally agreed responses impossible. Blair's call to established agreed principles to guide future interventions went unheeded.[2] With political agreement as difficult as ever to achieve, international legal sanction in the form of a UN Security Council resolution for military action in Kosovo, Sierra Leone, Iraq and Syria, among others, was not forthcoming.

Consequently, subjective justifications – emotive and individualised *ethical* claims – became increasingly important as governments tried to garner public support for military action to aid vulnerable individuals in far-flung parts of the world. Though these subject-oriented arguments can take many forms, human rights – and women's rights in particular – would play a significant and enduring role in trying to gain and sustain popular support for military interventions by the US, UK and other liberal Western governments. The wars in, and with, numerous states – Kosovo, Sierra Leone, Afghanistan, Iraq, Libya – were the continuation of multiple political dialogues that included the deployment of martial force. Associated with each military intervention are truth wars, which took place between competing moral frameworks and opposing political ideologies: some secular and some religious in origin. An important front in the intervention truth wars would endure throughout every military intervention: gender and the role of women in society.

In 1987 American political theorist Jean Bethke Elshtain wrote *Women and War*, examining how what she called the myths of Man as Just Warrior and Woman as Beautiful Soul 'helped to secure women's locations as non-combatants and men's as soldiers'.[3] She set out to expose the limitations of historically entrenched stereotypes by highlighting contradictions such as combative women and placatory men. In the 1990s the US, UK and their NATO partners advanced career opportunities for women within their respective armed forces. Air forces and navies provided new combat opportunities for women, while armies placed more women closer to – and occasionally in – combat and harm's way than ever before. It might be reasonable to expect, and many feminists did, that traditional stereotypes would begin to break down in the bloody business of war. In parallel, Joshua Goldstein incorporated approaches from several academic disciplines in exploring 'the main ways in which gender shapes war and war shapes gender' in a study that spanned the history of war.[4]

Yet for all the studies and efforts towards gender equality in civic life in the West, as well as in the armed forces, echoes of Elshtain's Beautiful Souls and Just Warriors continue to be found in the military intervention discourse of the twenty-first century: in justifying initial interventions, shaping the way they are fought, and in trying to sustain popular support for long-term military engagement. This chapter explores some of the ways in which gender, especially representations of women, has played and continues to play an important role in the justification and execution of recent military interventions, most notably in Afghanistan. Beginning with an examination of the portrayal of vulnerable, usually Muslim, women in Western media, the chapter will explore how governments use these representations to gain support for their preferred military actions. Discussion will then progress to some of the philosophical and religious underpinnings of these different perspectives on women, showing just how far back in history the basis of modern attitudes can be found. The chapter will conclude by assessing the extent to which, far from advancing the cause of women's rights around the world, ideologically-based gender claims by the US, UK and other Western allies can have the opposite and unintended effect of reinforcing long-standing cultural stereotypes.

Women in war

In the twenty-first century popular, long-established Western cultural images of the vulnerable, suffering woman in need of assistance from brave manly warriors have continued to be reinforced across multiple domains, from Marvel comic-based film franchises like Spiderman, Batman and Superman to serious Hollywood productions where lead characters continue to be predominantly male. When Western media speaks to the world from war zones like Afghanistan or Libya, or from the less violent but nonetheless dangerous uprisings in the Arab Spring, the woman as representative victim is still a favoured lens through which to constitute particular 'truths' about events on the ground. Whether it is about perpetuating myths, selling newspapers, contrasting unfamiliar cultures with our own in a less-than-favourable light, or attempting to confirm the superiority of equality-based democratic systems over religiously-driven patriarchies (or some combination of these possibilities), the vulnerability of women continues to shape – and be shaped in – competing discourses.

In the political arena, global leaders have regularly resorted to using the image of the vulnerable woman, set alongside equally vulnerable children, as shorthand for the desperation faced by entire communities and populations. From Blair to Obama, establishing and maintaining support for military interventions has been pursued in part by repeatedly stressing the need to protect women, often with the implication that women are more vulnerable than their husbands, other male family members, or wider communities. Examples from Tony Blair and George W. Bush include, from Kosovo, Afghanistan, Iraq and North Ossetia:

> The full extent of the horrific repression by Serb forces in Kosovo is only emerging now. There has been organised systematic rape of women, usually in front of husbands and children...The UNHCR today have substantiated reports that women and children are being used as human shields, including in a building used to store ammunition.[5]

> Our fight is with [the Taliban] regime, not with the people of Afghanistan. These people have also suffered for years: their rights abused, women's rights non-existent.[6]

I recall a few weeks ago talking to an Iraqi exile and saying to her that I understood how grim it must be under the lash of Saddam. 'But you don't,' she replied. 'You cannot. You do not know what it is like to live in perpetual fear.'[7]

This month in Beslan we saw, once again, how the terrorists measure their success – in the death of the innocent, and in the pain of grieving families. Svetlana Dzebisov was held hostage, along with her son and her nephew – her nephew did not survive.[8]

Similarly, when in 2013 Alex Thomson, television news correspondent, wrote and broadcast about the unfolding humanitarian catastrophe in Syria, he did so through the lens of Jannah Reid, a vulnerable woman trapped with her children in a dystopian nightmare come true.[9] Abandoned by her Syrian husband and struggling for food, the caring, nurturing mother fears for her children's lives and risks her own to try and ensure their survival: echoes of Elshtain's vulnerable, Beautiful Soul live on.

When wartime atrocities are given a face and a voice, they are frequently given the face and voice of a woman – preferably alongside that of a child: each reinforcing the vulnerability of the other. Krista Hunt explores the contradictions and conflicts in feminist responses to the war on terror. On one side the Feminist Majority Foundation (FMF) supported military action that would liberate Afghan women, while on the other, women's rights groups opposed the war as being against the interests of women.[10] Though I will not explore it further here, the war on terror opened up a new front in feminist truth wars.[11]

Elshtain's 2003 book, *Just War on Terror*, specifically promoted the need to protect the rights and lives of women in Afghanistan as a key justification for President Bush's military response to the 9/11 attacks on the US:

> Although pre-Taliban Afghanistan, a Muslim society, had included a significant number of professional women, women were forced under Taliban rule to withdraw from law, government, and teaching. These practices show us that gender practices are not a sidebar to the war on terrorism as a cultural struggle, but a central issue.[12]

More than a decade of bitter fighting later, with trillions of dollars spent and thousands of lives lost by NATO allies, gender politics in Afghanistan remain as much a fascination for Western observers as they were at the outset of the War on Terror. When moral comparisons of the place of women in the tribal Muslim cultures of the Afghan provinces of

Helmand, Herat and Kunduz (which have their own differences) and the place of women in the cultures of NATO members like the US, UK and Germany, the gap between them appears not to have narrowed.

The example of 15-year-old Sahar Gul highlights a gulf between Western perceptions of women's moral conduct and that expected from Afghan women: all of which takes place in relation to cultural, social and religious codes that have emerged over aeons. Sahar had been married to a 30-year-old man as part of an agreement between two families. At her new marital home she was tortured, starved and imprisoned in a locked cellar for months. Her injuries included burns, bruising to the face, hair torn out, and a fingernail pulled off. Somehow, the local police were made aware of Sahar's situation and rescued her from further torture.[13] So far, so familiar. The element of the story that highlights the gulf between Western moral sensitivities and the moral practices of day-to-day life for many Afghan women concerns the way in which Sahar achieved her freedom. Because she was rescued from captivity – as opposed to running away of her own accord – she could not be found guilty of a moral crime: unlike other Afghan women who have attempted to flee similar situations.

Such an attitude turns Western expectations on its head. Women who remain in abusive relationships, married or otherwise, are frequently judged in negative terms for not attempting to escape the domestic violence they endure – especially if children are involved – regardless of the moral repugnance or even depravity of the man inflicting violence upon her mind or body.[14] The *moral* thing for the abused woman to do is to escape. To display physical courage in the act of escaping is the 'right thing to do' and valued in its own right. When Amanda Berry risked recapture and further torture to flee from decade-long imprisonment, violence and degradation by Ariel Castro in May 2013 – saving her fellow abductees and captives Michelle Knight and Gina DeJesus in the process – three local Cleveland councillors set up the Cleveland Courage Fund[15] to raise money for the trio.

In contrast, women are frequently tainted by moral responsibility for their own victimhood in the West if they opt to stay and endure violence, coercion and intimidation, perhaps to the point of death. Newspaper headlines ask questions like 'Why did she stay?' when yet another dead woman's body is taken to the mortuary. It is therefore almost impossible to grasp that within the moral codes that shape and define Afghan culture and behavioural expectations (or perhaps more accurately, the moral codes that define the multiple, coexisting Afghan cultures), *leaving* is a moral crime on the woman's part. The final twist on this particular front in the gender truth wars is that the credit for rescuing Sahar is given to the police, the *men,* who rescued her. Even at her moment of freedom she remains a passive, vulnerable object whose liberty, such as it is, is delivered by men: there is no sense of the proactive striving or risk-taking bravery that characterised Amanda Berry's dash for freedom as she fled the Cleveland home that had been her prison. That is not to say that Sahar had never tried to escape: media reports state that she did.[16] Unfortunately for Sahar, neighbours returned her to her tormentors: tradition, community honour, obligation or a combination thereof shaping their actions and constructing an invisible prison in the process. These attitudes, however, did not arise overnight and echo the religious beliefs and ideologies that have shaped cultures and practices over time.

War, evil and women

A number of problematic philosophical underpinnings can be found in some of the justifications for military intervention put forward by George W. Bush and Tony Blair – especially where their ideas are applied to issues of equality and the protection of women. For example, when ancient Western just war concepts are set alongside notions of human rights as defined by the more recent 1948 UN Universal Declaration. Consider first the context in which Augustine's just war emerged 1,600 years ago.

When, in the fifth century, Augustine contributed to the just war tradition by writing about war and just war, he did so at a time of multiple and overlapping forms of violence. The Roman Empire was under sustained attack like never before in its 1,000-year history.[17] The Catholic Church was fighting to maintain religious dominance within the Empire in the face of internal disputes over what counted as 'true' doctrine, as well as demands by some to once again recognise the ancient Roman gods who had successfully protected Rome in the past. Augustine, the dominant Christian theologian of his age – perhaps any age – was cautious about the use of military force and when it could be used. He expressed his justification of war primarily in terms of the political entity of the empire, state or city and cast it in a mainly defensive or reactive light. He echoed Cicero who declared that a city should not wage war unless it is defending either the faith of its people or its physical safety.[18] Augustine also wrote that just wars are those which avenge previous injury by an enemy, the subsequent failure of that enemy to punish culpable citizens, or to regain property that was wrongfully seized.[19]

When these ancient ideas are applied to Afghanistan in justifying the 2001 war in response to the Al-Qaeda attacks on the US, there is a degree of historical resonance. The US had been attacked and claimed reactive self-defence as its primary justification of the use of force. Although, strictly speaking, it was not the ruling Afghan Taliban that orchestrated the attacks on American soil, they hosted the Al-Qaeda faction that did. Compounding their culpability in US eyes, they refused to hand over Osama bin Laden and his followers to the American authorities after the attacks took place. However, using ancient just war ideas in any justification of military action in the twenty-first century requires selective use of those historical concepts: problematic elements need to be overlooked or left safely in the past. For example, neither George Bush nor Tony Blair would have strengthened their arguments abroad, especially in the Muslim world, if they stated that their justification of military intervention relied on just war ideas that emerged in a strongly Christian context. An excerpt from Augustine's writings demonstrate both the advantages and disadvantages of his early just war approach:

> The desire for harming, the cruelty of revenge, the restless and implacable mind, the savageness of revolting, the lust for dominating, and similar things – these are what are justly blamed in wars. Often, so that such things might also be justly punished, certain wars that must be waged against the violence of those resisting are commanded by God or some other legitimate ruler and are undertaken by the good.[20]

The idea of just punishment of the cruel as a response to 9/11, or savage acts against innocent victims in third-party countries, might have broad appeal in parts of the US and beyond. However, the 'just punishment' approach has long been prohibited in International Law by the UN Charter, which only allows war on the basis of either national self-defence – individual or collective self-defence – or by Security Council agreement to restore international peace and security.[21] Even more problematic is the reference to a command by God, with inevitable associations with a form of holy war.

Throughout their respective periods in office, a degree of controversy surrounded both Bush's and Blair's personal beliefs and motivations, with religious imagery and ideas appearing regularly in their intervention discourse. Bush was less guarded than Blair on this point. On 16 September 2001 the President framed his, and America's, response to the attacks that took place five days before in overtly religious tones: 'This is a new kind of – a new kind of evil. And we understand. And the

American people are beginning to understand. This crusade, this war on terrorism is going to take a while.'[22] While the semantics of what he meant by 'evil' can be debated across religious and cultural divides, there is enough common meaning in the word for his listeners to know he was saying something more than just: 'bad people have done a bad thing'. When the word 'crusade' is added to the equation – a word that would surely not have appeared in a scripted speech – the potential for alarm and offence in Muslim countries increased hugely, given the historical associations of the word: regardless of whether or not he was merely seeking to reassure the American people at a time of national crisis. More subtly but with similar undertones, four days later Bush stood before a Joint Session of Congress and spoke simultaneously to the American people and a worldwide audience. His challenge was simple, 'Either you are with us or you are with the terrorists,' words easily recognisable to his domestic Christian audience as an adaption of Jesus' words in Matthew's gospel: 'He who is not with me is against me'.[23]

These and other religious references added credence to comments in 2005 by Nabil Shaath, Palestinian foreign minister, at a 2003 Israeli–Palestinian summit in Egypt attended by President Bush, who claimed: 'President Bush said to all of us: "I am driven with a mission from God". God would tell me, "George go and fight these terrorists in Afghanistan". And I did. And then God would tell me "George, go and end the tyranny in Iraq". And I did.'[24] Although the White House subsequently denied the words attributed to the President, in the intervention truth wars factual accuracy counts for little when the stakes are high and war is on the horizon. It almost does not matter whether Bush said these words exactly as quoted, whether he said something similar to those words but was mistranslated, or whether it is a complete fabrication – intended or otherwise. Given his other utterances, the perceived truth of that particular incident rests on one's view of President Bush, his government's interventionist policies at that time, and social media's ability to send the story round the world and back before officials could confirm or deny the claims. The reported claims were, for many, absolutely real.

In 2006, with the Afghanistan and Iraq wars raging and insurgent violence killing more civilians than soldiers along social and religious divides, Blair publicly expressed his belief that God – amongst others – would judge his actions:

> That decision has to be taken and has to be lived with, and in the end there is a judgement that, well, I think if you have faith about these things then you realise that judgement is made by other people,

and ... if you believe in God, it's made by God as well ... but in the end you do what you think is the right thing.[25]

Blair's decision to invade Afghanistan and Iraq and conduct interventions elsewhere – like his moral decision making generally – was informed by his Christian faith which would, in turn, lead to judgement in the afterlife. Their personal beliefs and the way that both Bush and Blair sought to present themselves as ethical in their political actions were all informed by multiple truths: religious and ideological. The most basic claims to religious truth in the Christian tradition are that God exists, is knowable and reveals his will to people who believe in Him; and that such religious claims really are 'true', by implication subverting competing truth claims. While subsequent leaders in the US and UK in particular have avoided the overt religiosity of Bush and Blair, claims about the role of women in society, equality, freedom and democracy have consistently contributed to intervention discourses.

In writing about the US-led NATO military action in Afghanistan, Elshtain uses Augustine and his early Christian ideas to set out a philosophical and theological basis for the relationship between men and women. [26] Put simply, Western gender practices are to be encouraged, while Taliban-enforced cultural practices are repressive, degrading and violent towards women. Part of Elshtain's claim is that the ethical individual cannot and should not seek to opt out of tough decisions in the political realm, or support those who do.[27] Specifically, her pro-interventionist stance fuels her demand for the United States to engage positively with the world, resorting to violence in the pursuit of justice where necessary. For her, ethical leaders make tough decisions and ethical observers lend their support: it is not just about following one law or being limited by another law, but deciding on how one human or group will act in relation to another human or group. She states: 'It is unsurprising that we flinch and heartening that so often human beings rise to the occasion as they answer generously and forthrightly not only the question, Who am I? But also, Who is my neighbor?'[28] Recognising oneself as ethical (and recognising one's enemy as unethical), and using that self-recognition as a prompt to help others goes to the heart of intervention discourse.

Excerpts from the speeches of Blair and Bush in 2001 and 2004 point to progress in Afghan public life away from the unethical dictates of the Taliban towards the liberal freedoms they have both espoused and fought for. Blair described the consequences of failing to confront the ideological and religious enemies of not only the West but of moderate Muslims as well:

> What bin Laden and the Taliban regime represent is a form of Islam that is not shared by the vast majority of Muslims. Certainly any Muslims I have spoken to. And secondly, that what they want to see is Taliban-type regimes all over the Middle East and elsewhere. And those regimes are repressive, dictatorial, brutal in their treatment of women, and usually regimes that end up being economically disastrous as well.[29]

Three years later, and claiming some form of shift towards the freedom and democracy he was seeking for Afghanistan, Bush told the United Nations:

> The Afghan people [have] adopted a constitution that protects the rights of all, while honouring their nation's most cherished traditions. More than 10 million Afghan citizens – over 4 million of them women – are now registered to vote in next month's presidential election.[30]

Bush's optimism seems misplaced a decade and several interventions later. The ugliest and probably most accurate truth to emerge from a conflict, whose desired strategic outcome has changed with the political winds, is that the West is tired of military intervention. Any pretence at an idealistic ongoing commitment to Afghanistan's future has been replaced with hard-nosed pragmatism as the US, UK and their NATO partners attempt to withdraw their troops by the end of 2014. The rhetoric of women's rights, democracy and equality that has been a consistent feature of the justification of ongoing military action is now being marginalised. This is not a new development, however, the portents for women in Afghanistan have been made clear for several years.

In April 2010, Abdul Hadi Arghandiwal, the Hezb-i-Islami political faction leader, was reported by Human Rights Watch as telling a group of leading Afghan women that they would have to 'sacrifice their interests for the sake of peace' with the Taliban.[31] Whatever deep-rooted liberal democratic ideals Bush, Obama, Blair, Cameron and other Western leaders have drawn upon to suggest that they were fighting for women's equality, have been set aside. When, in August 2010, Time magazine published a photo of 18-year-old Afghan woman Aesha Mohammadzai on its cover, the inference was that such women needed to be protected and that the fight against oppression must go on.[32] Aesha, like Sahar Gul, had been abused and tortured – in this case for years instead of months. She escaped her abusers but was subsequently caught and jailed for five months for the crime of running away.[33] When a judge finally returned her to the husband she fled, he and other family members cut off her ears and nose as punishment for the 'dishonour' she had brought upon them. The published photographs remain as a visual representation of the cultural and ethical fault-line that still lies between Afghan and Western treatment of women.

While American plastic surgeons have managed to reconstruct Aesha's mutilated face, the underlying cultural mores that allowed it to happen – and the religious beliefs that underpin them – continue unchanged. In March 2012, Afghanistan's Ulema [religious] Council published a declaration enshrining the rights and responsibilities of women based on a strict reading of Sharia law. President Karzai endorsed the declaration – and its philosophical basis in Islamic belief – that officially confirmed women's subordinate position in Afghanistan, both in the privacy of the household and in public society: 'In consideration of the clarity of verses 1 and 34 of Surah an-Nisa' [of the Qur'an], men are fundamental and women are secondary; also, lineage is derived from the man'.[34] Trillions of dollars, thousands of lost lives and a decade of Western liberal democratic-motivated military intervention would appear not to have resulted in any permanent freedoms for Afghan women and girls. A final insult – if that is possible – is that the declaration was made and endorsed by President Karzai while Hilary Clinton, the most powerful woman in America at the time – served as US Secretary of State.

Mixed messages

There are few more vulnerable and provocative statements in the political realm than the ideological truth claim, of any hue. Every such truth necessarily provokes and represses opposition. Consequently, attempts to use gender in the discourse of war and military intervention ultimately expose a number of ideological contradictions between Islamic views (either moderate or extreme) on women's place in society and the secular, idealised liberal Western view of equality, and Western Christian understandings of the relationship between men and women (these last two should not be taken to be the same thing). They also expose a number of mixed messages on gender, political contradictions and practical failures in all of them. Take three examples: the signal failure of Christianity and the Church – all churches combined rather than a specific denomination – to live up to its own gender ideals; the gulf between Western liberal idealism and the everyday experiences of women; and the ignominy of 'giving' equality through war.

The founding document of Christianity – the New Testament – has shaped religious, cultural, social and political thought since it was written almost 2,000 years ago. In contains an ideal – articulated by St Paul – that offered a radical alternative to the subsequent experience of women and men in the Christianised West in particular. Paul wrote of the essential equality of all people before God: 'There is neither Jew nor Greek, slave nor free, male nor female, for you are all one in Jesus Christ.'[35] Setting aside the complexities of the day-to-day practices of those to whom St Paul was writing, and the dangers of reading a snippet of religious text as a stand-alone piece of wisdom, there is a spectacular failure at the heart of this Christian claim. The Church has never truly believed it or implemented it: at least not in any way that is recognisable to twenty-first century liberal notions of equality in an era of universal human rights.

Despite St Paul's biblical injunction about the unity and equality of all Christian believers, accepted social and cultural gender practices of the day played the dominant role in shaping both the early formation of the Church and its subsequent development. From the Eastern Orthodox churches to the Roman Catholic Church and more recent Protestant denominations that are almost beyond number, the overall ratio of men to women in leadership positions is skewed massively towards men.[36] This dynamic is usually reinforced by traditionalists with reference back to Jesus' gender and the fact he chose 12 men to follow him as disciples. All manner of intellectual contortions are made to somehow justify this contradiction: 'equal but different'; 'value is not linked with role': Exactly the same kinds of justification produced by Muslims to explain the relationship between men and women. A key difference between Christian and Islamic attitudes to gender emerges from the modern separation of religious and civic life for Christians in a way that is not countenanced by many Muslims.

Moving from the concept of equality in the Church to the notion of protecting the weak, the Church experiences its greatest failure and greatest shame. The sexual and physical abuse of children by clergy – compounded by institutional cover-ups across denominations – has been increasingly exposed in the latter decades of the twentieth century and the opening years of the twenty-first, stripping away Christian moral authority in the public sphere. In a public hearing on child abuse in Geneva in January 2014, the Vatican was accused by a UN panel of violating the 1990 UN Convention on the Rights of the Child by moving abusive priests from diocese to diocese rather than reporting them to police, and 'sweeping offences under the carpet.'[37]

When it comes to equality and fairness for women, Western liberal idealism has fared somewhat better than the Church – though it too should not be held up as an unequivocal success. In historical terms, voting rights for women, even in the West, is a relatively recent development. Although the Constitution of the United States of America – the founding document of US law and the principles upon which American democracy rests – was adopted on 17 September 1787,[38] it did not grant equal voting rights to all Americans. It took until 1920 and the 19th Amendment to the US Constitution to bring about universal suffrage, granting all American women the right to vote. In the UK, most women over 30 were enfranchised in 1918, though it took until 1928 for equal voting rights with men to come into existence.

There are British and American women alive today whose mothers were not allowed to vote and whose lives were blighted by the kinds of violence and degradation seen in the many accounts of Afghan women published during the ongoing war. Worse, there are gender practices taking place in the US and UK that match many of the horrific excesses of Afghan brutalising of women. In 2013, the UK's Royal College of Midwives estimated that there were 66,000 victims of Female Genital Mutilation (FGM) in the UK, with a further 23,000 girls 'from African communities' under the age of 15 'at risk of – or may have undergone – FGM'.[39] Similarly, in the US, as far back as 1997 it was reported that an estimated 168,000 women and girls were either living with FGM or were at risk from FGM.[40] It is entirely possible that a girl born in the US to parents of Somali, or other African, origin could grow up to be the President of the United States. Statistically, however, it is much more likely that she will be pinned down by a close relative and, without anaesthetic, have her clitoris cut off. These cultural practices may have migrated to the West but it appears that liberal values like the principles of equality and the force of law are insufficient to protect the bodies – and minds – of these young, now Western, girls. It is a sad truth that in countries where gender has been used as part of the justification for military intervention, women and girls are at risk – on a large scale – of a practice that falls within the UN definition of 'torture or other cruel, inhuman or degrading treatment'.[41]

In one war-related area NATO countries have had a measure of success in advancing equality for women: their introduction to military roles that have been the exclusive domain of men for decades, sometimes centuries. President Obama and Prime Minister David Cameron, like

President Bush and Prime Minister Blair before them, take every opportunity to remind their countries and the world that American and British women have been actively involved as combatants in recent wars:

> The character of our military through history – the daring of Normandy, the fierce courage of Iwo Jima, the decency and idealism that turned enemies into allies – is fully present in this generation. When Iraqi civilians looked into the faces of our servicemen and women, they saw strength, and kindness, and good will.[42]

> I want to begin by paying tribute to our men and women in uniform who, once again, have acted with courage, professionalism and patriotism.[43]

> I know what side of history I want the United States of America to be on. We are ready to meet tomorrow's challenges with you – firm in the belief that all men and women are in fact created equal, each individual possessed with a dignity that cannot be denied.[44]

> For 12 years now, men and women from all parts of these islands have been serving their country in Afghanistan ... Sacrifice beyond measure – from the finest and bravest armed forces in the world. And I want us to stand, to raise the roof in here, to show just how proud of those men and women we are.[45]

In these and many more examples the leaders concerned go to great lengths to identify women explicitly in the achievements of their respective armed forces. Grammatically, there is no reason to refer to those achievements and sacrifices made by 'men' and 'women'; generic phrases such as 'military personnel' or 'soldiers' would enable listeners or readers to know who is being referred to in the speeches. The explicit inclusion of 'women' enables them to be associated with activities and characteristics historically associated with men: battle, sacrifice, victory, bravery, patriotism. These women are no longer Elshtain's 'Beautiful Souls' but have become 'Just Warriors' alongside men: demonstrating the freedom, equality, and dignity associated with the liberal Western democracy that is being encouraged in – or forced upon, depending on your perspective – repressive Muslim countries. Medal and awards ceremonies reinforce the 'warrior woman' effect and provide a rare opportunity for women to work and compete on equal military terms with men. There is some way to go before equality on parts of the battlefield is matched by equality off it. Given the choice between a man who has won a medal for bravery and a woman who has won the same medal for a similar action, Western media frequently focuses on the latter. Similarly, when a woman has been killed in Iraq or Afghanistan it remains a story, long after the press has moved on from the deaths of their male counterparts. The most prominent 'woman-at-war' story to emerge from recent US military interventions, however, is one of notoriety and national discomfort.

Equality is not quite what it appears when the warrior woman violates the ideals she is supposed to represent. In lectures in several different institutions and countries – including the US and UK – I have asked audiences to name anyone involved in the torture or abuse of Iraqi prisoners at Abu Ghraib prison. With one exception, the first, and sometimes only, name to be offered has been that of Lynndie England, a US Army Reserve Specialist in the 372nd Military Police Company. England, and the prisoner abuse more generally, can be viewed in different ways. For example, Bruce Tucker argues that the individualisation of the Abu Ghraib violence – blaming a small number of mainly junior personnel – allowed Americans to avoid facing up to issues such as racialised violence and the wider moral questions surrounding the war on terror.[46] At the level of the subject, another dynamic emerges: gender difference and the unspoken but all-too-present cultural and moral expectations that persist.

In Western armies, in operational theatres like Iraq and Afghanistan, women – enlisted and officers – can be found close to, and sometimes at, whatever passes for the front line in contemporary war fighting. This proximity to bullets and roadside bombs is incidental to some medical and administrative roles. Women are generally still not appointed in their own right to 'teeth' arms where the ability to inflict overwhelming, lethal violence is their *raison d'être*: the infantry, the marines, and the Special Forces. Woman as 'just warrior' is still in the shadows of her killing counterpart. Yet at a time when the status of just warrior has not been fully realised for fighting women, Lynndie England's actions represent another step further away from the vulnerable, fragile Beautiful Soul who needs to be protected: she has become its nemesis.

England is singled out because the protector (sic) became not a just warrior but an unjust abuser: an abuser of prisoners and an abuser of the cultural expectations laid upon her. She became the face of cultural difference and anti-American, anti-Western propaganda on every front where the US and its allies were fighting. A decadent product of a decadent society who wielded physical power over Muslim men, humiliated them merely by observing their nakedness, further violated them by making them simulate, or carry out, sexually degrading acts, all compounded by the use of dogs – considered filthy by many in that region. Even in these most extreme of actions England could not find equality with her co-conspirators – her actions somehow represented as worse, more culpable. She is an embodied contradiction of the aims of her President, government and Army to extend democracy, equality and freedom. England's actions are also an affront to women in Afghanistan, whose hopes and lives are often as vulnerable as the prisoners at Abu Ghraib but whose imprisonment is ensured not by bricks and barbed wire but under the unrelenting gaze of custom, culture and religious sensibility, reinforced by the inescapable threat of violence.

Summary

On 8 October 2001, the day after American and British missiles were launched against Afghanistan's ruling Taliban regime in response to the Al-Qaeda attacks a month before, Tony Blair spoke of protecting civilians, especially women: 'They are victims of the Taliban regime. They live in poverty, repressed viciously, women denied even the most basic human rights and subject to a crude form of theocratic dictatorship that is as cruel as it is arbitrary.[47] Like the concern for the women with which this chapter started and ends, concern for women provides the emotional and ethical book-ends for NATO's military involvement in that benighted land. The hubris of NATO's leaders is captured in Blair's words after the so-called defeat of the Taliban only a few weeks later in November 2001: 'Although conflict is never easy or pleasant, to see women and children smiling after years under one of the most brutal and oppressive regimes in the world is finally to understand the true meaning of the word "liberation".'[48]

How much are those women and children still smiling as NATO forces withdraw from that country? How much have they smiled as a resurgent Taliban has pursued its bloody insurgency against what, for them, are foreign occupiers and their illegitimate puppets in Kabul? Gender politics has been a constant presence in the intervention truth wars, in Afghanistan and elsewhere. However, like the women whose rights are to be sacrificed as some form of accommodation is made between the Afghanistan government and the Taliban, the politics of equality are being set aside in London and Washington as ugly political truths reassert themselves. There are no votes to be found in foreign graveyards; there are even fewer votes to be found among the flag-draped coffins that are borne with grief-laden dignity as their occupants make a final journey to the lands from which they came. Worse, the very armies that have gone to Afghanistan as liberators are unable to uphold their own ideals within their own ranks. Women continue to be regularly subjected to bullying, harassment and sexual assault in the US, UK and other armed forces. In 2009, the US Department of Defence reported that around a third of women in the military had been sexually assaulted,[49] a statistic that remains stubbornly high several year later.[50] In the UK, similar concerns over the treatment of women in the army have been expressed by Madeleine Moon MP, a senior member of the UK Defence Select Committee, not only about the number and severity of bullying and sexual harassment incidents but also the Ministry of Defence's failure 'to take the issue seriously enough' and to tackle it with sufficient rigour.[51] If women in Afghanistan, Iraq, Libya, Syria and beyond are to enjoy – if that is the word – the freedoms that women are claimed to have in Western liberal democracies, much more blood will have to be spilt. Such freedoms – if they ever come – will have been bought in blood and pain by those who seek them if they are to be worth anything. They cannot be given or implemented, not even by armies containing women.

6
Drone Wars

These events actually happened. On 27 March 2013 a hostile drone[1] could clearly be seen slowly circling, its pilot preparing to launch the first ever drone attack against an American military installation. Security personnel at Creech Air Force base watched in fascination as the drone sought out its target, eventually striking down seven civilians. There was no panic, no screaming or shouting, no wailing sirens from rapidly approaching fire trucks, only solemn silence as the prone bodies were outlined in chalk.

A continent away at RAF Waddington in England, one of two bases in the world from where the Royal Air Force operates the MQ-9 Reaper drone over Afghanistan, a similar attack was imminent. The controllers were tense as the missile trajectory indicated an impending, precision strike on the gate that could be clearly seen on the screen.

'Go straight,' urged a well-meaning observer.

'Keep going,' added one of the men whose foreheads glistened with the strain of responsibility and the physical effort required.

As the missile struck its target with a clean hit, everyone standing close enough to see the action cheered. Then confusion reigned. 'Now what?' seemed to be the unspoken question as the unexploded warhead remained in its final position, a testament to those whose combined actions had brought about this successful 'hit'.

Unusually for drone strikes, cameras were at hand to capture every detail on film. Footage of the strike at Creech Air Force Base, Nevada would go straight to YouTube,[2] while the second strike on the UK would be broadcast by the BBC later that day. These were no ordinary images and these were no ordinary drone strikes. They were not carried out by an enemy state and they were not launched from the air. They were both simulated drone strikes by anti-drone groups who were protesting against ongoing drone operations from Creech Air Force Base, and the start of UK-based control of drone operations over Afghanistan from RAF Waddington in Lincolnshire. The missiles involved were neither the 100lb AGM-114 Hellfire air-to-ground weapon of choice for precision strikes against the Taliban,[3] nor its larger, more destructive 500lb sibling, the GBU-12 Pave way. No, the missile that appeared on British TV screens was around eight feet long, one foot in diameter and borne aloft to its target on the shoulders of six protestors. The phallic symbolism would be amusing if the point being made was not so serious. The American protestors did not even replicate a missile strike, they relied on a mocked-up Reaper to strike and 'kill' its civilian targets directly by touching them.

What prompted these and similar protests to be made in the US and the UK?[4] Why did RAF Waddington attract the most high profile peace camp to be set up in the UK since the Greenham Common Women's Peace Camp protested against the presence of American nuclear weapons in the 1980s? It is difficult to find the answer in statistics alone. American and British governments have been reluctant to publish statistics on drone killings, either the deaths of intended targets or the unintended collateral deaths of innocent civilians. President Obama has acknowledged the dilemma: 'There's a wide gap between US assessments of such casualties and non-governmental reports. Nevertheless, it is a hard fact that US strikes have resulted in civilian casualties, a risk that exists in every war.'[5] Accurate statistics are notoriously difficult to acquire and are inevitably contested. The UK Ministry of Defence publicly announced that on 25 March 2011 a Royal Air Force Reaper operation in Helmand Province, though successful in killing two insurgents and destroying their explosives, had resulted in the deaths of four civilians.[6] In contrast, the claimed figures by non-governmental agencies like Reprieve and the Bureau of Investigative Journalism (BIJ) for civilian casualties of US drone operations in Pakistan are much higher.

A major non-governmental report claimed that between 2004 and 2012 around 3,000 people were killed in drone strikes in Pakistan, with almost a thousand of those being civilians.[7] Based on a cold calculus of killing, some of the concerns of the most strident anti-drone campaigners appear disproportionate when these casualty rates – each one a tragic loss for the families and communities concerned – are viewed in light of events elsewhere and possible alternative weapons. The Reaper and the Predator are, fully armed, capable of far less destruction than their manned-aircraft counterparts, their maximum payload weight of almost

4,000lbs dwarfed by the 70,000lbs carried by the nuclear-capable B-52 bomber. Even if the reported total ten-year US drone death figure for civilians in Pakistan of almost 1,000 is accurate, it is dwarfed by the 130,000+ killed in the Syrian civil war (at time of writing) where conventional hand-held battlefield weapons are the primary killing tools. Worse, 1,000 deaths equates to about two hours' killing in the Rwanda massacre of 1994 where almost a million people were killed in three months: mainly by machete and garden tools.[8] Significantly for the politics of drone warfare, polls carried out in the US and UK in 2013 show strong public support for the use of drone strikes against terrorist targets.[9]

The remainder of the chapter will, first, explore how Predator and Reaper drones with lethal strike capability – as opposed to drones whose use is limited to intelligence-gathering or surveillance – came to occupy such a ubiquitous place in the Afghanistan War and the wider War on Terror from Pakistan to Yemen and Somalia. In the process, three distinct applications of drone capability will be considered in the operations of three separate institutions: the CIA, The United States Air Force (USAF) and the Royal Air Force (RAF). Then second, the arguments of opponents and proponents of the technology will be examined, looking at how subjective, individualised considerations are framed by competing truth claims about Reaper and Predator in contested political, military and morality discourses: all with the aim of influencing the behaviour of observers and gaining their support.

Rise of the deadly drones

For two decades, technical advances in computing, optical and communication technologies, the emergence of terrorist threats and attacks at home and abroad, added to the global defence industry's insatiable appetite for profits and new markets, have all combined to fuel a relentless, exponential proliferation of drones. The toy remote controlled aircraft of bygone days has undergone multiple military makeovers, resulting in a range of surveillance platforms and weapons systems that is so diverse and ubiquitous that many academics and public commentators have gone so far as to claim that the very nature of war has changed forever.[10] (There is no time to discuss these claims here so I will merely observe that similar statements have been made throughout history about the changing nature of war in response to the advent of the spear, the longbow, the rifle, the artillery piece, powered flight, nuclear weapons and so on. All of these technological changes are still absorbed by Clausewitz's maxim that the essence of war is a violent clash of wills.)

Another less obvious factor began fuelling the rise of the drone long before the Twin Towers were reduced to smouldering rubble of twisted steel, acrid dust and unrecognisable human remains. US military intervention in Somalia in the early 1990s reached its nadir in 1993 in the Battle of Mogadishu – more famously known as Black Hawk Down. Captured US soldiers were dragged through the streets of the town, killed and mutilated. The senseless barbarity of the violence prompted Americans, from Main Street to the White House, to question the value of sacrificing young soldiers in lands and for causes that were so far removed from the United States and its founding ideals. Dying for a noble cause in defence of one's homeland and family is one thing; dying for an obscure cause on behalf of people who express no gratitude and who display no sympathy for the liberal democratic values that motivated US intervention in a tortured land is quite another.

By the time it became clear that the war against Al-Qaeda would be far more complex and time-consuming than the toppling of the Taliban regime in Afghanistan in October and November 2001, and that the search for Osama bin Laden had turned into a cross-border manhunt across Afghanistan and Pakistan, the Predator drone had developed – driven by the CIA as well as the United States Air Force – from an advanced unmanned reconnaissance and surveillance system to a platform capable of extended observation time and the delivery of precise, lethal force. In parallel, as the Iraq War degenerated into a full-scale insurgency involving competing domestic interests and growing resentment of foreign occupiers, the Predator could be deployed in yet another operational theatre.

By 2004–2005, the interventions in Afghanistan and Iraq had turned into a mixture of terrorism, counter-insurgency, police action, fragile nation-building and a resurgence in the poppy growth that fed the addiction of Western heroin addicts and kept Taliban fighters supplied with money and weapons. Concurrently, the drone became politically irresistible and opened up a new front in military intervention and its associated truth wars: multiple capabilities (reconnaissance, surveillance, intelligence-gathering and precise, lethal firepower) with no risk to the crew involved. The loss of a Predator would not be accompanied by a loss of American life. In the politics of military intervention, the transfer of physical risk from aircrew overhead to civilian bystanders on the ground held great appeal in Washington and, later, London. In the battle for the hearts and minds of the American people, and likewise elsewhere among NATO allies, every flag-draped coffin chips away at the public support base on which – in democracies at least – the political will to continue fighting rests.

Drones offered Presidents Bush and Obama a tool with which to mount unconventional offensive action in places like the Taliban-controlled region of Pakistan where a dead or captured American pilot – or worse, a captured pilot killed on camera for the benefit of the radical Islamist and the internet voyeur – would be politically calamitous and utterly unacceptable on Capitol Hill. Consequently, a linked but apparently semi-detached CIA anti-terror drone campaign escalated in Pakistan, Yemen and Somalia as the Bush and Obama administrations progressed. If figures from the Bureau of Investigative Journalism are even close to accurate, by the start of 2014 President Obama had authorised almost seven times more drone strikes than his predecessor: 330 to 51.[11]

Problematically for the US government, an increase in the number of Taliban and Al-Qaeda fighters being killed by drone strikes in Pakistan, was accompanied by an increasing number of civilian casualties. The numbers of Pakistani non-combatant deaths is hotly disputed by both sides, though some attacks have been severe enough to prompt the Pakistan government to close key NATO supply routes between Afghanistan and Pakistan. The inadvertent killing of 28 Pakistani soldiers in a US drone strike in November 2011 provoked political, military and popular outrage across Pakistan, throwing an international spotlight on US cross-border activities.[12] This and similar events over several years – attacks on wedding parties, signature strikes that operate on the basis of guilt by profiling and by personal association – has prompted a powerful counter-discourse on drones that challenges the claims and the ethics of the US government.

As the use of drones has proliferated in the United States Air Force, the CIA and, later, the Royal Air Force, political opposition and anti-drone campaigns have grown in intensity from Pakistan to the US and the UK. Claims that drones offer a humane, precise means of striking an elusive enemy are rejected by opponents who point instead to the civilian collateral damage involved and the violation of Pakistan's state sovereignty, preferring the representation of remote killing as a dishonourable, cowardly means of inflicting death on people who cannot fight back. One consistent element of the drone truth wars is the subjective focus by opponents on the individual: the distant 'killers' who deal death from afar; 'innocent' bystanders whose lives are sacrificed in a futile cause; and the implied 'noble' Taliban and Al-Qaeda fighters who are fighting an asymmetric war against an overwhelming enemy. In parallel, the American and British governments have displayed a consistent unwillingness to discuss drone operations, perhaps hoping that a 'see-no-evil, hear-no-evil, speak-no-evil' approach would lead to critical voices just fading away. Reinforcing this perception, senior administration figures from both countries have regularly visited Afghanistan for photo opportunities with 'battle-hardened troops' who put their lives on the line against an elusive enemy. Meanwhile, no American president or British prime minister has published a photo of themselves on a morale-boosting visit to a Reaper or Predator squadron.

Not surprisingly, and often highly creatively, Taliban and Al-Qaeda fighters have modified their tactics to maximise their advantages while striking at their enemy's points of vulnerability, one of which is their enemy's determination to act within military law and the Geneva Conventions. Increasingly sophisticated improvised explosive devices (IEDs) – roadside bombs, land mines, booby traps – have taken a huge physical and psychological toll on ISAF (International Security Assistance Force)[13] foot patrols. When combined with shoot-and-run sniper ambushes of US and UK personnel, the use of IEDs has become even more effective for the insurgents. Embracing the publicity benefits of showing the world – supporters and potential supporters in particular – how the anti-ISAF campaign is succeeding, Taliban and Al-Qaeda attacks are videoed on mobile phone cameras and shared on the internet and other forms of social media: ambushes, IEDs, sniper attacks and sometimes a combination of the three. Complicating the situation further, the UN has recognised that increasingly sophisticated, industrial-grade weapons material that originates outside Afghanistan is being deployed against ISAF forces.[14] Unfortunately for local Afghan civilians, the Taliban's roadside bombs and booby traps are entirely autonomous and indiscriminate, as likely to kill local civilians as the enemy troops they are targeting.

Yet the use of roadside bombs and IEDs in Afghanistan does not attract the same degree of international opprobrium as the deployment of drones, with Special Rapporteur Christof Heyns invoking strongly anti-drone discourse in his 2013 report to the UN.[15]

The US and UK administrations have been reticent in engaging the full range of government machinery in contesting the drone truth wars, preferring to play down the use of drones in the hope of assuaging growing disquiet about their use. Part of the difficulty in making a moral assessment of drone use is that even within the US the same equipment – Predators and Reapers – is used in different ways by the US Air Force and the CIA: the former employing military rules of engagement and the latter, given the nature of covert CIA-led operations, being used in a non-transparent and, in the eyes of opponents, unaccountable way. The picture was made even more complicated when, in March 2013, US Central Command announced that it would no longer publish monthly data on drone strikes in Afghanistan because critics were 'disproportionately focused' on the use of weapons in strike operations (three per cent of sorties) in relation to their much wider reconnaissance and surveillance tasks (97 per cent of sorties).[16]

The politics of drone killings

Recognising the increasingly controversial nature of drone operations President Obama addressed the matter directly in a major speech in May 2013. He outlined the American position:

> Beyond the Afghan theatre, we only target Al-Qaeda and its associated forces. And even then, the use of drones is heavily constrained. America does not take strikes when we have the ability to capture individual terrorists; our preference is always to detain, interrogate, and prosecute. America cannot take strikes wherever we choose; our actions are bound by consultations with partners, and respect for state sovereignty. America does not take strikes to punish individuals; we act against terrorists who pose a continuing and imminent threat to the American people, and when there are no other governments capable of effectively addressing the threat. And before any strike is taken, there must be near-certainty that no civilians will be killed or injured – the highest standard we can set.[17]

Despite these remarks, opponents perceive that the US is acting with impunity, advancing its own defence and political interests at the expense of innocent collateral damage to victims abroad. Some would argue on human rights grounds that even confirmed terrorists should not be killed in this way, even when no practical means of arrest or detainment exist, or that in the attempt many more unintended deaths could occur. Balancing strong domestic support for the US drone policy with a concern for human rights and political transparency, Obama even announced that he had asked his 'administration to review proposals to extend oversight of lethal actions outside of warzones that go beyond our reporting to Congress.'[18] He refers here to strikes in countries like Yemen and Somalia where no sovereign state apparatus exists that can deal with the threats to others that emanate from within their borders.

Across the Atlantic, the Royal Air Force's use of Reaper in Afghanistan causes even greater domestic controversy than does its American counterpart – this despite only four confirmed collateral damage deaths by drone strike.[19] The contrast between UK and US figures for civilian deaths as a consequence of lethal drone strikes has become a source of dispute.

Drone Wars UK,[20] part of a British anti-drone alliance has questioned the accuracy and, by implication, the honesty of the UK government's claim that only four civilians have been killed as a result of drone strikes in Afghanistan, because the number is so much lower than the claimed figure of almost 1,000 for CIA drone operations in Pakistan.[21] The lack of public engagement by the UK Foreign Office and Ministry of Defence is frequently presumed in anti-drone discourse to mean that bad things are happening but are being hidden by bad people from the public gaze of concerned, good people. An alternative but less intriguing explanation is that only four civilian collateral damage deaths by UK drones have actually occurred.

There are a number of possible reasons for the statistical disparity between the claimed US-caused and UK-caused deaths of civilians by drones. The British government and the Royal Air Force could, as opponents imply, be lying. Myriad civilian deaths might have happened and they somehow 'got away with it'. The cost of being caught out misleading the public, however, would be far higher than admitting to a collateral death in the first place. The PR advantage to the Taliban and other opposition groups would be huge and they have every incentive to produce evidence that supports claims against the British and Americans. Despite this possibility, the most likely explanation for the difference between RAF and CIA figures for civilian deaths lies in the rules of engagement that were set by their respective governments. Neither the American nor British governments publicise the specific rules of engagement to which its drones are operated. Despite this, there is a clear basis for difference between the approaches of the two allies. The US is operating wartime rules of engagement: it continues to target Al-Qaeda elements in Pakistan, Yemen and beyond (on top of military operations in Afghanistan) as part of its response to the 9/11 attacks. If there are differences between USAF and CIA rules of engagement these remain confidential but are set in the context of a country at war against Al-Qaeda and its allies.

The British position is significantly different. Having not been attacked on 9/11 – the legal and moral basis for the Afghanistan War and the American War on Terror – the UK is not at war in the same way as the US. As a result, its rules of engagement are less permissive, essentially being based on the principle of self-defence. Limitations on British operations were reinforced by a judgement on 19 June 2013 by the UK Supreme Court, confirming that British military personnel are subject to the Human Rights Act (1998), which in turn enshrines the European Convention on Human Rights (ECHR) in British law: even beyond the geographical confines of the UK and continental Europe. The judgement 'requires the state not to take life without justification and also, by implication, to establish a framework of laws, precautions, procedures and means of enforcement which will, to the greatest extent reasonably practicable, protect life.'[22] Related to this approach to engaging the enemy, personal discussions with American former Predator pilots in 2013 highlighted occasional tensions between the allies when it came to defending troops under attack on the ground. British drone crews were not being authorised or willing to strike when there was any possibility of civilian deaths – even with increased risk to American, British or other allied soldiers on the ground. This should not be seen as evidence of greater moral concern by the UK and its drone operators over their American counterparts. Instead it should be understood as a reflection of the political contexts within which each operates. Further, my ongoing research with the UK's two Reaper squadrons – including visits to both at Creech Air Force Base, Nevada and RAF Waddington, England – highlighted both personal and institutional heightened awareness of government-level political sensitivity to the deaths of civilians.

Legality and morality

The moral basis of the American response to the attacks of 9/11 was part brilliant, part problematic, depending on the perspective of the observer. On 20 September 2001, President Bush declared the War on Terror and identified the enemy who would be pursued to the ends of the Earth: 'Our war on terror begins with Al-Qaeda, but it does not end there. It will not end until every terrorist group of global reach has been found, stopped and defeated.'[23] From a non-American perspective, one problematic, legalistic reading of Bush's declaration rests on his lack of a state entity as the target for his war, which was motivated partly by self-defence, partly by reprisal. Afghanistan's then ruling Taliban regime was to be (and subsequently was) targeted if it refused to hand over Osama bin Laden and his Al-Qaeda co-conspirators to the United States. This despite no direct act of war being waged by the state of Afghanistan against the United States of America. Article 51 of the UN Charter – written and signed by the US at the end of World War II – acknowledges that every state has the right to self-defence in international law, though the Charter is framed in terms of state actions rather than the actions of sub-state groups.[24] For non-Americans who did not endure the gut-wrenching reality of being part of a nation under attack, the legality of launching a full-scale war against a state in response to an attack by a terrorist organisation remained hazy. Despite this, sympathy for the plight of a country traumatised by the sight of thousands of citizens being atomised in a cloud of smoke, fire, and dust prompted many international onlookers not to ask too many awkward questions. Since, for Americans, Congress authorises war and the Supreme Court is the ultimate arbiter in law, distant voices quibbling over the niceties of international law and the more problematic elements of the UN Charter were not ignored – they were not even heard.

Ethically speaking, however, the just war tradition[25] offered President Bush considerable flexibility of response. Though increasingly juristic in recent centuries, just war is not framed exclusively in legal terms. The right of self-defence is not necessarily predicated on an attack by an equivalent power or political entity (the Roman Empire was not under attack by another empire at the start of the fifth century: Rome was attacked by Alaric and his army of Visigoths from the north). The brilliance of the response by the Bush administration was in setting out its *justum bellum* in conceptual terms – the War on Terror – rather than in geo-political terms. Brilliant, because the War on Terror could be whatever the President wanted it to be, even if it opened him up to the humourist's charge that he had declared war on an abstract noun. The approach was still somewhat problematic, however, because such a broad definition exposed the US to the charge of writing a blank cheque to itself in terms of who could be attacked and on what grounds. The subsequent, tenuous linking of Saddam Hussein and Iraq to the War on Terror further damaged the credibility of the original declaration: Saddam had no sympathy for, and shared no ideology with, bin Laden and Al-Qaeda. As a result, a War on Terror that could potentially be against anything and everything could not be sustained in the long-term: philosophically, financially or politically.

By 2013, President Obama was keen to place some spatial and conceptual limitations on America's ongoing war against those who had attacked it in the past and who threatened it in the present and future:

> Beyond Afghanistan, we must define our effort not as a boundless 'global war on terror,' but rather as a series of persistent, targeted efforts to dismantle specific networks of violent extremists that threaten America...Moreover, America's actions are legal. We were attacked on 9/11. Within a week, Congress overwhelmingly authorized the use of force. Under domestic law, and international law, the United States is at war with Al-Qaeda, the Taliban, and their associated forces. We are at war with an organization that right now would kill as many Americans as they could if we did not stop them first. So

this is a just war – a war waged proportionally, in last resort, and in self-defence.[26]

President Obama linked America's ongoing drone campaigns – including military action in Afghanistan and CIA covert action in Pakistan, Yemen and Somalia – directly to the events of 9/11. In doing so he used the same just war terminology as his predecessor, language that can be traced back many centuries.[27] Combined with the legal sanction provided by Congress in response to the 9/11 attacks, these just war criteria provide the codes within which the US continues to operate more than a decade later. Despite restating the American case for using drones against its enemies in terms of legal and moral codes, contesting voices continue to grow louder. Problematically for both Bush and Obama, those voices have tended not to focus on the overarching moral and legal arguments, disputing the detail of just war thinking or the constitutional basis for taking a country to war. Opposition has been increasingly framed in terms of the individual subject: the people who suffer from drone attacks and the distant personnel who carry out those attacks.

Innocent victims and immoral killers

The drone truth wars are therefore predominantly framed in terms of the individuals at both ends of the drone 'kill chain': operators in an air conditioned cubicle at one end and, more than a continent away, dead innocent civilians at the other.[28] For some reason the cubicles being air conditioned is a consistent part of drone discourse, emphasised by opponents to highlight the contrast between the relative physical comfort of Predator and Reaper crews and the hot, dusty discomfort experienced by their targets. The discomfort of the personnel inside the theatres of operations whose job it is to get the drones safely into the air and back down again at the beginning and end of each sortie is overlooked. (Remote crew based in the US and UK only operate the Predators and Reapers once they are safely in the air and hand control back to locally-based operators when it is time for the aircraft to land.)

The most popular representation of drone pilots and sensor operators[29] by opponents of the technology is the 'Playstation killer'.[30] This is the apparently physically and emotionally distant, desensitised young military operator who is unable to distinguish between video games and real war, who sees targets not as human beings but as blips on screens to be indiscriminately eradicated. These assumptions have been repeated so often and by so many commentators that they have gone from being contested truth claims to assumed fact. In *Wired for War*, Peter Singer writes about the relationship between risk and bravery among those who operate drones thousands of miles from war zones. He considers the drone pilot to be the first combatant in history to break the link between soldiers and the warrior values that have defined them through the ages. For Singer, the drone crew is 'disconnected' from the war that they pursue.[31]

Physically speaking, Singer's argument appears entirely logical and plausible: how can someone whose greatest risk is a combination of boredom and repetitive strain injury be classed as a warrior? If an enemy managed to shoot down a Predator or Reaper, or if mechanical failure caused one to fall from the sky, the only indication to the crew would be a lack of control response and a camera image of the ground rushing towards them, followed by a blank screen as communication with the stricken drone ceased. Singer's view is reinforced by Peter Olsthoorn who writes about the psychological distancing – paralleling the physical distancing – of drone crews from their targets. He concludes that killing Afghans, or anyone else, in such a fashion could become significantly easier from the safety of a distant control room thousands of miles away in Nevada.[32] Likewise, Medea Benjamin promulgates the assumption that killing from a distance becomes 'easier'; remote operators will not have to endure the daily dread of the roadside bomb or sniper attack that awaits patrols on the ground.[33] (Which is, as advocates point out, the whole point of drones.) Like Singer's claims, Olsthoorn's and Benjamin's arguments somehow seem intuitively 'true', their truth reinforced and accentuated by uncritical repetition across multiple media outlets by journalists who, like some authors whose theory and supposition is presented unquestioningly as fact, have never been near an operational drone squadron. However, no matter how logical or plausible it sounds, unquestioned, untested truth is little more than folklore or legend.

A research visit to Creech Air Force base to spend time talking to Reaper crews and watch them in action provided a contrasting perspective. On the matter of it being 'easier' to kill from a drone as has been suggested by many commentators, a Reaper pilot shared his experience:[34]

I have killed the enemy from both [conventional aircraft] and from the Reaper. The body's reactions are the same – it surprised me. Your mouth goes dry and the hairs on the back of your neck stand up. Everything goes tense and you get that sick feeling in your stomach. You know what you are about to do.[35]

The experience of one or even several Reaper crew members does not speak for everyone involved, but the words of this pilot are given credence by the findings of ongoing research. Almost half of all USAF Reaper and Predator crew members surveyed reported significant levels of stress, with conducting combat operations being one of the stress factors. Further, Reaper and Predator pilots are displaying similar rates of post-traumatic stress disorder (PTSD) as their counterparts conducting operations in conventional manned-aircraft.[36] Furthermore, one of the stressors of conducting drone operations, faced by both USAF and RAF personnel, comes from the pressures, institutional and personal, *not* to kill civilians near targets. The following incident involving a British Reaper crew – which echoes sentiments shared with me by a small number of American Predator pilots – highlights the challenges of decision making in potentially life or death situations:

> I've had multiple strikes where waiting a little longer, or using the extra situational awareness tools in the Reaper have resulted in much better outcomes than you'd have got from a manned-aircraft in the same setup. It happens almost every day. My last flight involved working with [soldiers on the ground], who wanted us to provide some ISR[37] on a hotly-contested area where they encounter a lot of IEDs[38] and a lot of sporadic, harassing fire. We saw, before sunrise, a man leave a compound and go to an area behind a building. He started digging, interacting with the ground. The controller [on the ground] saw that and immediately suggested that it was an IED, and started trying to arrange permission for us to strike under the 'hostile act' ROE.[39] His thinking was that there was a recent IED strike nearby, it was suspiciously before sunrise, and this was near an entryway to the compound, so probably a defensive IED.
>
> My crew disagreed, and as we watched longer and more closely, we could pick out some of the tools he was using and started to assess them as regular farming tools. Eventually with the first fringes of sunrise, we could tell he was just seeding a small patch of ground. Watching him for an hour let us see that he had none of the hallmarks of a traditional IED emplacer; there was no rapidity, no hurry, no equipment, no lookout...I had a team inside the ops room I could talk to at length on the phone; second/third/fourth opinions available as required, and a feed that could be instantly stopped, rewound and reviewed to gather more information. A manned-aircraft with less equipment, less time, and a poorer camera would have almost certainly considered engaging [weapons earlier].[40]

The controller on the ground in Afghanistan had, based on recent similar experience, a reasonable basis for commencing the process for a drone strike on what he saw as a dangerous target. If the popular representation of the drone pilot as the disconnected Playstation killer was even vaguely accurate – and the policy of the government and air force involved was apathetic to the loss of civilian life – there was no discernible reason for any delay in blowing up the suspected bomber. It would have been 'easy'. Further, instead of the potential target being a mere blip on the screen to be eradicated, the resolution of the picture enabled the Reaper crew to assess the man's actions on the basis of *identifying the farm tools he was using*. The farmer was left to go about his business unharmed. The assessment of the Reaper crew member offers a final irony: that a manned aircraft – which does not share the popular opprobrium heaped on the Reaper and Predator – would have been more likely to engage and kill the potential target.

Summary

It would be naïve to assume that either opponents or advocates of drone operations have a monopoly on truth in their claim and counter-claim. The politics of military intervention continues to redefine the parameters of the drone wars in a number of ways. In the US, President Obama has responded to concerns by exploring means by which greater scrutiny of drone operations outside of war zones – Yemen, Somalia, Pakistan – can be achieved. Despite strong, ongoing domestic support for the use of drone strikes as part of American defence and security policy, the Obama administration, the CIA and the US Air Force have clearly taken steps to ensure a reduction in civilian deaths in Pakistan at least. At the start of 2014 the Bureau of Investigative Journalism, arch-critic of drone strikes, reported that in 2013, for the first time since drone operations began against targets in Pakistan, no confirmed civilian casualties were reported.[41] Meanwhile in the UK, political sensitivity to civilian deaths caused by such an unpopular means of delivering air power has engendered a near obsessive determination on the part of the RAF and its two Reaper squadrons to avoid civilian casualties in Afghanistan at almost any cost – even when they are providing air cover to allied forces on the ground. Four deaths in 2011 remain the Ministry of Defence's only acknowledged collateral damage statistic and no evidence has been put forward to support a claim to the contrary.

Civilian deaths by drone continue to be reported in Yemen and Somalia, despite President Obama's determination to minimise the innocent casualties of war. And here is the dilemma that will continue to drive the drone wars and the truth war within it: Should the United States aim for zero civilian casualties even if it will reduce the effectiveness of its anti-terror(ism) campaign? Even the Geneva Conventions do not demand, or provide for, complete immunity for civilians in military operations (recall, the US is still conducting a 9/11-induced self-defence war against those who would attack it). Article 51 of the 1977 Protocol I Additional to the Geneva Conventions describes as indiscriminate, 'an attack which may be expected to cause incidental loss of civilian life, injury to civilians, damage to civilian objects, or a combination thereof, which would be excessive in relation to the concrete and direct military advantage anticipated.'[42] Unpalatable though it is for many, it is not against international humanitarian law to conduct operations in which civilians die, as long as there is a direct military advantage to be gained. What is considered a direct military advantage remains, of course, under dispute.

However, in the drone truth wars it is the very use of weapons systems that are piloted from the other side of the world that appears to be most unacceptable to critics. Practically speaking, if the problem is the use of drones then it could be solved overnight given the resources of the American military in particular. Conventional aircraft could provide the air-to-ground attack capability necessary to strike identified Al-Qaeda targets in places like Yemen and Somalia. B-2 or B-52 bombers could be deployed to devastating effect, though they would more than likely destroy entire towns as kill individual targets. Probable result: more civilian deaths. Alternatively, soldiers – Special Forces – could be air-dropped in foreign lands to kill or capture identified targets. The likely result is more military *and* civilian deaths. Perversely, an anti-drone discourse that is predicated on preserving the lives of small numbers of civilians would, if taken to one logical conclusion, actually cause greater numbers of civilian deaths through the deployment of more powerful, conventional, military alternatives.

Another possibility that emerges from taking anti-drone discourse to a second logical conclusion is that Predator and Reaper are not the real focus of their vocal opponents but serve as a kind of discursive lightning rod for anti-war activities in general and anti-US military interventions (even when these are undertaken alongside British and other allies) in particular. Morally, such intentions are laudable – so long as everyone agrees that war, terrorism and insurgency should henceforth become a form of non-contact sport. Furthermore, unilaterally adopting non-violence as a government policy in a still-violent world would mean that the American President, like the leaders of other drone-capable states, would have to abrogate responsibility for defending his people from attack. The ultimate consequence of taking anti-drone claims to such a logical conclusion would be more dead civilians overseas in the short term and, as a result of surrendering the right to use a precise and relatively moderate force in self-defence against enemies, more dead civilians at home in the long-term.

Part III
Politics, Truth and the Global Financial Crisis

7

It's All Your Fault

By 2006 one financial truth dominated the world's advanced economies: the good times are here to stay. In November that year, Ben Bernanke, Chairman of the US Federal Reserve, showed no concern as he looked to the future:

> I have confidence, therefore, that however events play out in the short term, in the longer term the economy will grow at a healthy pace, raising living standards in the process. The Federal Reserve will continue to play its part by implementing policies designed to achieve its mandate of fostering price stability and maximum sustainable employment.[1]

Similar sentiments were being uttered at the financial centres of all major economies as success begat success: bankers' bonuses erupted, financial institutions grew at unprecedented rates, tax revenues were rising and confidence was soaring. It would later become politically convenient for the rest of the world to point the finger of blame at America for what would happen in the years to follow, but the same lack of concern was endemic. The hubris of the time was perhaps articulated best by Gordon Brown, who could never be accused of false modesty when it came to his complete and utter command of his finance-brief as the British Chancellor of the Exchequer. Should anyone in the UK or beyond be unsure of his financial genius he subsequently made clear the extent to which he and his government had triumphed over the capricious whims of the economic cycle:

> 'I have said before: no return to boom and bust.'[2] (22 March 2006, Budget Speech)

> 'And we will never return to the old boom and bust.'[3] (21 March 2007, Budget Speech)

The benefits of Brown's bullish approach – loved by his party and supporters as financial genius at work and only half-heartedly derided by his fearful, admiring opponents – were felt at the ballot box during successive General Election victories (1997, 2001, 2005) as the economic bubble was inflated by soaring debt levels. Brown was not alone in his optimism, delusion as it turns out, with political leaders, finance ministers, economists and financiers around the globe basking in the glory of a period of almost unparalleled growth. Yet within two years the global economic landscape would go from El Dorado to post-apocalyptic. The four trillion dollar question came from an unlikely source during a visit to the London School of Economics in November 2008 when Queen Elizabeth II, voicing the concerns of millions of bemused, confused and angry people, asked: 'If these things were so large, how come everyone missed them?'[4]

Countless books, newspaper and journal articles, documentaries and crisis meetings, and national and international commissions have sought to answer variations on this question. Complicating matters further, the subjective aspect of the financial crisis and resultant economic recession takes multiple forms. Bankers, hedge fund managers, insurance brokers, traders in futures and other highly complex products have been collectively constituted as incompetent, immoral, and probable law-breakers. Their contributions will be analysed in the next chapter, 'Governing Greed'. Meanwhile, this chapter will proceed by setting the financial crisis in context, outlining key factors that contributed to the global financial market coming perilously close to melt-down, with the collapse of major banking and other financial institutions under the burden of unsecured and uninsured, or uninsurable, debt. The remainder of the chapter will explore the deeply personalised discourse surrounding two specific groups, political leaders and subprime borrowers: showing how the identities, attitudes and behaviours of these two specific target groups were shaped in the global game of blame and counter-blame.

Distant warnings

In 2008, former US Federal Reserve Chairman Alan Greenspan offered some early reflections on the financial calamity that was unfolding before the eyes of the world:

> In recent decades, a vast risk management and pricing system has evolved, combining the best insights of mathematicians and finance experts supported by major advances in computer and communications technology. A Nobel Prize was awarded for the discovery of the pricing

model that underpins much of the advance in derivates markets. This modern risk management paradigm held sway for decades. The whole intellectual edifice, however, collapsed in the summer of last year because the data inputted into the risk management models generally covered only the past two decades, a period of euphoria.[5]

This brief excerpt from a longer statement indicates a number of factors that contributed to a regime of truth about the state of financial markets and the unwarranted degree of trust people had in them. Computers and communications technology are widely regarded as unquestionable goods in the advancement of the human race, while mathematics is viewed in almost mystical terms both by its greatest proponents and by those who understand it the least. With mathematics providing the basis of scientific, architectural and technological advances from the days of Pythagoras and even before, the very term engenders a degree of respect and admiration – perhaps especially among those for whom statistics, probability and the use of integrals to calculate the area under a curve are alien concepts. Further, there are so few people capable of critiquing the validity of mathematical financial models at the level of a Nobel Prize winner – which in itself connotes respectability and credibility – that errors in the models might (and did) go undetected until it is too late.

One of the difficulties is that few outside a university mathematics department are able to tell their proofs from their theorems, or their conjectures from their models. And sometimes people who do understand the differences between them obfuscate and mislead the less mathematically informed because it suits their purposes to do so. The force of deductive reasoning in mathematical proofs is reflected in the certainty that a particular proof, once demonstrated, will always be true in all circumstances. That is the kind of certainty financiers and politicians like.

The Royal Swedish Academy of Sciences 1997 award of the Bank of Sweden Prize in Economic Sciences in Memory of Alfred Nobel is described in its press release as follows: 'Robert C. Merton and Myron S. Scholes have, in collaboration with the late Fischer Black, developed a pioneering formula for the valuation of stock options. Their methodology has paved the way for economic valuations in many areas. It has also generated new types of financial instruments and facilitated more efficient risk management in society.'[6] However, when Alan Greenspan referred to the Nobel Prize, there was no mention of the limitations of mathematics-based financial modelling generally, only the acknowledgement that risk management models had been populated with inadequate data that did not cover a sufficiently long period.

More significantly, the very codes, laws, rules that were meant to safe-guard the financial system and the interests of everyone linked to it – and which would shape ethical conduct in the field – were founded on flawed assumptions, erroneous calculations and perverted perceptions of risk. Certainty levels were over-claimed, confidence grew artificially high and poor decisions were made as a result.

In 2009 Charles Bean, Bank of England Deputy Governor for Monetary Policy, addressed the Annual Congress of the European Economic Association, outlining in retrospect why the financial crisis occurred and then developed into a subsequent global recession which few countries escaped. His speech captured the shock and surprise felt by economists around the world, as well as the crystal clear hindsight with which they reviewed recent events:[7]

> Low interest rates and low apparent risk created strong incentives for financial institutions to become highly geared. Unfortunately much of this leverage occurred off-balance sheet in order to avoid on-balance sheet capital charges. The innovative financial instruments developed to achieve this were highly complex. This complexity did not seem to matter when markets were steady and defaults low but was fatal once conditions deteriorated, as it became impossible to understand and price these instruments objectively. Leveraged institutions were trapped; as their assets sank in value and funding dried up, buyers became wary – a classic lemons problem. As losses moved inexorably towards institutions at the heart of the financial system, the complexity of the inter-bank network created enormous uncertainty about the extent of counterparty risk.[8]

The way in which Bean set out his analysis, using what he refers to as 'standard economic models', demonstrates the impenetrable vocabulary that has left the lay observer wondering, first, if anyone actually knew what was happening at the time, and second, if anyone understands even now. Bean even asked a rhetorical question of his audience in Barcelona: 'Wasn't it blindingly obvious that the whole house of cards would come crashing down at some point?'[9] The short answer was: yes. However, it only seemed obvious to a small minority of economists and politicians whose influence was not enough to dent the dominant regime of financial truth.

One such Wise Owl, as Bean would later describe them, was Nouriel Roubani. In 2006 he predicted to an audience at an International Monetary Fund (IMF) seminar that the combination of three factors would lead the

US economy into a major slowdown, followed by a recession. The first was an impending housing bust – not a slump or a slow landing but a 'bust'; the second was an energy price shock based on major oil price increases; while the third factor was an impending delayed reaction to interest rate rises by the Federal Reserve.[10] His analysis was much more in-depth, of course, but a key strand of his argument was a comparison of the situation in 2006 with the recession of 2001, highlighting significant commonalities between events surrounding an impending housing bust with the high tech bust of 2000–2001. In explaining how he came to his conclusions Roubani said: 'My analysis has been based on circumstantial kind of observations…So my model is like a 'smell test' or a 'duck test': if it looks like a recession and walks like a recession and quacks like a recession, it should be a recession.'[11] Regardless of the detail of his analysis and ominous forebodings, when he suggested that the probability of a recession was 70 per cent, referring in the process to circumstantial evidence and including the 'duck test' in his reasoning, it is not difficult to understand why his audience did not leap to their feet in panic. On the contrary, when he concluded his speech the Moderator Charles Collyns quipped: 'Thanks very much, Nouriel. I think perhaps we will need a stiff drink after that.'[12] The audience laughed. Further, before the crisis, 'wise' was not what Roubani and others who shared his views were called. Roubani's counsel was as welcome as that of the religious prophets of old, with their dire warnings, counter-cultural ways and frequently shortened lifespans. His negative predictions earned him the nicknames Dr Doom and 'permabear'.[13]

So what took Roubani from being the derided Dr Doom to a Wise Owl in two years? Events. Bad things happened, mainly in line with the key dangers as he predicted. Exactly what happened, how and why will be argued over for decades – ultimately without agreement between the contending interest groups. And how, you may ask, can I as someone who is not an economist (but who does have a grasp of applied mathematics that will exceed that of many economists, politicians and bankers) make such a confident assertion? Because the key protagonists in the financial truth wars have interests and reputations to protect, and any unfavourable judgement on their activities will leave a stain that time will not erode. In addition, the number and complexity of the overlapping causes of the crisis brings into conflict a series of truth claims that cannot in any objective way be measured against one another. For example, a political desire for higher tax receipts cannot be measured against the limitations of a mathematical modelling technique; bankers' bonuses cannot be related directly to physical productivity; domestic financial rules have limited effectiveness in a globalised economy; the ideological underpinnings of American capitalism and Chinese capitalism are ultimately incommensurable; social factors impinge in often unacknowledged and even unseen ways; to name but a few.

A major foundation stone in the financial edifice that eventually crumbled was the idea that risk could be accurately measured and a price could be applied to it. Then, the value of risk became a commodity in itself: bought and sold like other futures trades. Complicating matters further, Alan Greenspan could later refer to his 2005 warnings about the 'protracted period of under-pricing of risk',[14] yet as Chairman of the Federal Reserve and one of the most powerful financiers in the world he was unable to stop the practice. If risk management was one of the factors that made the financial crisis possible, when other factors are set alongside it the challenge of finding a definitive explanation increases exponentially. These factors include, but are not limited to: a major credit bubble in the US, Europe and beyond; a housing bubble in many major economies; macroeconomic trends that were poorly understood, with previous crashes being assiduously ignored; ratings agencies whose perceived credibility hugely outweighed their ability to accurately analyse the complex situations and products they were commenting upon; rapid financial sector growth where risks were magnified at a time when they appeared to be managed under 'light-touch' regulatory regimes; burgeoning debt – personal, institutional and governmental – to GDP ratios; creeping leverage levels in what would become the most calamitous game of 'dare' in history; insufficient securitization of loans; unlimited trading – gambling – on the futures markets; imbalanced trade deficits that resulted in some countries having too much cash and needing to get a return on it, with other countries cash-poor and desperate to borrow at low interest rates; failure of corporate governance – too much focus on the bottom line and not enough awareness of the dangers involved; and flawed economic and financial theories.

An official US report into the cause of the crash was published in 2011: *The Financial Crisis Inquiry Report: Final Report of the National Commission on the Causes of the Financial and Economic Crisis in the United States*.[15] It addresses, sometimes using different terminology, all of the above factors, and more, that contributed to the crash. Given that the Commission was instituted by Congress and signed into being by President Obama in May 2009 with ten experts appointed by Democratic leaders and Republican leaders from the House and the Senate, it might be assumed by the unwary that it would produce a definitive version of events and decisions that led to the financial crisis.[16]

The report summary likened events to a car crash on a highway that had no speed limits or painted white lines within which to order the behaviour of the drivers. In the vocabulary of this book, most of the people involved in making key decisions on the long build-up to the crash did not have the option of conducting themselves ethically by conforming to established rules, laws and regulations. For the most part there were, in practice, no effective codes. Where codes did exist there was erosion over time of the constraints they provided, compounded by an increasing unwillingness to enforce them, such as they were. Subjective ethics, or lack of ethics in the pursuit of personal self-interest, played an increasingly important role. Analogies of a financial Wild West – not limited to the United States, though the sentiment at the heart of the analogy still applies – are possibly more accurate than the example of the naked highway, which at least had the travellers on a sound stretch of tarmac (asphalt).

The report reached nine main conclusions. *The financial crisis was avoidable* – human conduct featured at every stage as subprime lending and securitisation soared, mortgage debt and house prices exploded, and regulation of the mortgage market failed. *Failures in financial regulation and supervision destabilized US* [and global] *financial markets* – lack of political will rather than lack of authority was identified as the main culprit. *Failures of corporate governance, especially regarding risk management* – reckless short-termism ultimately overwhelmed capital balance sheets. *Excessive borrowing, high-risk investments, and hidden leverage levels* – from household budgets to major financial institutions. *Lack of government understanding of the financial system as it went into melt-down, and an uncertain response when realisation dawned. Lack of accountability and a failure of ethics. Panic spread rapidly as mortgage-lending and securitisation standards reached crisis levels. Derivatives like credit default swaps fuelled the crisis. Credit ratings agencies exacerbated the growing problems as, overwhelmingly, their ratings turned out not to be worth the paper they were written on.*[17]

Ultimately, what the Commission produced was a report that could serve as a single-document case study in the politics of financial truth. Despite every member of the Commission having access to the same evidence, the report was not unanimously agreed upon. A Commission which purported to be independent – made up of private citizens with expertise that ranged from economics to housing, finance, banking and more – separated along the partisan lines of their nominators when the report was eventually published. The four Republican-nominated members added their own dissenting statements to the report: Bill Thomas, Keith Hennessey and Douglas Holtz-Eakin collectively writing one alternative conclusion, with Peter Wallison providing another. The official report and the dissenting statement of Thomas, Hennessey and Holtz-Eakin both reflect the ideological positions and attitudes of the respective members to free-marketeering, regulation and enterprise.

Wallison's dissention is in some ways the most interesting in terms of demonstrating competing truth claims. His key criticism of the official report is that it spreads blame too evenly over the nine areas outlined above: if everything is blamed then nothing is held responsible. Wallison allocates blame most prominently – though not exclusively – to the US government's housing policy from 1992 onwards, which, he argues, ultimately led to 27 million subprime mortgages at the start of the financial crisis: a disaster waiting to happen.[18] Claims and counter-claims continue around Wallison's position.[19] Howard Davies recognises this argument and puts forward an alternative explanation: that the problem was not too much government interference (through housing policy), but too little government intervention and a lack of financial regulation.[20] Regardless of the accuracy or otherwise of Wallison's argument, what he illustrates clearly is the interplay of politics, truth and ethics in the matter of public policy. What he does more forcefully than any of his colleagues on the Commission is place prime responsibility at the feet of successive governments and their housing policy of increased home ownership: a policy specifically aimed at expanding home ownership among African-American, Hispanic and other low-income groups. He then argues that mortgage underwriting standards were eroded over time, both in the government sponsored enterprises Fannie May and Freddie Mac, which were used to facilitate housing policy, and in the corporate lending sector.

Here is one of the fronts in this particular battle in the financial truth wars. Which is the greater or most important truth: the ideological pursuit of greater social opportunity for a marginalised sector of the population, or the technical analysis of default rates, loan-to-income ratios, and other measurements of financial health? There is no objective way of making that judgement. Wallison claims that the Commission lacked objectivity from the beginning, meaning that a bi-partisan agreement on the facts could not be reached.[21] The philo-sophical problem he unwittingly identifies – which was echoed (though with different outcomes) by his colleagues on the Commission – is his foundational assumption that there is an objective truth to be reached in the first place. It is epistemologically impossible for someone whose personal politics and ethics prioritise greater distribution of wealth and social opportunity to see the 'truth' about the crisis in the same way as someone who believes that markets should be self-regulating, with survival of the fittest determining the bottom line.

Truth claims made on either side of this divide are infused with ideological assumptions, personal ethical stances, power politics, party-political agendas and other factors besides. The technical impenetrability of these various truth claims and counter-claims ensures that the non-specialist becomes quickly excluded from the national or international conversation. If financial 'experts' did not fully understand the risk assessment models that they used in contributing to the crisis, lay observers from either Capitol Hill, Urban Central or Main Street, Suburbia are even less likely to understand them. Political truth as objective reality is a mirage. The subjective alternative, however, is much more appealing; it is readily understandable and is premised on a very simple question: Who is to blame?

The blame game

Every child, especially those with siblings, learns at a very young age the best ways to avoid blame for any accident or wilful damage: deny all knowledge; claim to have been elsewhere; put someone else in the firing line; reject all evidence of personal involvement, regardless of how damning it is; form alliances of blame against unwitting, innocent third parties; blame unlikely freak natural occurrences and ruptures in the laws of physics; repeat all of the above until a guilty verdict is pronounced and punishment meted out. Then continue as before: it might work next time. Revisiting these guidelines that served me so well as a young boy with two brothers it occurs to me that while most people grow out of such behaviour, or at least the worst extremes, it is ideal preparation for a career in low politics or high finance.

An early offering in the financial blame game was provided by Davies, whose book entitled *The Financial Crisis: Who is to Blame?* gets straight to the point.[22] To some extent his approach mirrors the American Financial Crisis Inquiry Commission, though in a more digestible format and with less obvious partisanship (though he admits to his work being 'opinionated'[23]). More focused, and narrow, in their analysis were Andrew Hindmoor and Allan McConnell who asked of political leaders, treasury officials and regulators: 'Why Didn't They See it Coming? Warning Signs, Acceptable Risks and the Global Financial Crisis'.[24] However, rather than attempt to find the *real, true* or *objective* reason for the financial melt-down – which I have already argued is not possible anyway – the remainder of the chapter will look at two broad groups and examine how subjective positions were shaped by competing truth claims, individual ethical choices, inadequate moral codes and poor decision making as the crisis hit.

First to be considered are political leaders and financiers involved at governmental policy level. Ben Bernanke and Gordon Brown did not between them create the regime of financial truth that was set out at the beginning of this chapter. Though it seems now that the 'no return to boom and bust' attitude was both ridiculous and hubristic, Brown was only annunciating what many political and financial sector leaders privately thought from Washington to Berlin to Tokyo. Second, there is the 'ordinary citizen', the person who was foolish enough to believe politicians, gullible enough to apply for loans and mortgages into which they were being aggressively tempted but could not conceivably repay, and naïve enough to think that somehow things would turn out all right in the end. But, they were encouraged to do so by real estate agents who were frequently willing to be complicit in the completing of inaccurate, sometimes downright fraudulent, and increasingly foolish mortgages and other loans. In other words, the poor were blamed for being poor, constituted as irresponsible, described as a burden, and characterised as lazy and feckless.

Politicians and finance

Political leaders – and I am chiefly concerned with leaders in Western liberal democracies, however loosely that term is understood – form their identities, behaviours, attitudes and ambitions in relation to multiple overlapping, sometimes contradictory, ideological, social, party-political and personal factors. At the same time, those same political leaders – their identity, conduct and views – are also *shaped by* other competing forces in public discourse: political opponents who would undermine them at any turn; political *allies* who would undermine them at any turn; local constituencies with narrow interests and who make or break careers; multiple media outlets that shape public perception but over which there is limited or no control, to name but a few. Complicating matters, the moral contradictions and pragmatic constraints faced by political leaders mean that they have much less room for personal manoeuvre, or 'wriggle room', when it comes to decision making than is often recognised by those who criticise them. Vested interests that might have negative or destructive consequences elsewhere on the political landscape are never far away. Consider the following questions as the practical implications of the subjective elements of the 'political life' are analysed further with respect to the financial crisis.

First, what is the subjective 'stuff' on which the identity, behaviour, attitude and ambition of a political leader is built and where is truth located within it? The number of potentially influential factors is almost unlimited but can include concepts as diverse as ideology, religion, morality, philosophical view of the world, a desire to do good, a will to power, narcissism, altruism, personal ambition, and self-advancement. Each of these factors has its own relationship with truth and therefore a unique relationship with the financial truth wars.[25] Take ideology and politics, which have been inextricably linked since the term 'ideology' was first coined by Antoine Louis Claude, Comte Destutt de Tracy in 1796.

Ideologies are more than simply ideas, political or otherwise. Ideologies invoke whole frameworks of practices alongside their conceptual foundations: shaping language (look at the range of responses prompted by the phrases 'political correctness' or 'climate change'), political groupings and public policy, and challenging an existing *status quo*. Consequently, a politician whose fundamental personal belief or ideological persuasion is that workers should be freed from the tyranny of a ruling elite and be allowed a 'fair share' of collective wealth will see the financial crisis differently to someone who prioritises the individuality of *laissez-faire* economics and its acceptance of wide divisions between rich and poor. Furthermore, each will claim their ideological position as somehow 'right' or 'true' in such a way that will inform other decisions they make. So the financial truth wars begin (though of course do not end) at a highly individualistic, subjective level.

Second, what motivates politicians to recognise obligations towards others? Political leaders do not move in a vacuum, completely disconnected from what is called 'the real world' around them – though it is a common criticism they face. More accurately, their 'reality' is constructed from many factors that do not overlap with the day-to-day reality of their poorest constituents. Behaviour in office – positive, negative or politically neutral – can result from multiple stimuli: a desire to do good; will to power; preserving national heritage and prestige; rejecting heritage in favour of progressive ideals; making the world a 'better' place; shaping the world in my image; party loyalty; sexual, alcoholic, narcotic or some other compulsion; and religious discipline. Most of these stimuli contain some form of truth or morality claim, overt or otherwise. Doing 'good' contains an inherently moral claim (though what counts as 'good' is still contested), whereas will to power suggests the opposite – self-interest manifested in a desire to wield power over others. Further, religious discipline can be prompted by acceptance of specific truths. For example, the Christian who is motivated to advance the cause of the poor in response to some form of divine imperative will also, by definition, accept (usually as a faith-act) doctrinal statements about the nature of God and the associated divine moral order.

Third, how are political leaders prompted, coerced or otherwise persuaded to behave ethically (or unethically)? At this point the tensions between the various truth discourses that shape personal conduct can become problematic. A democratically elected politician might find herself in a moral paradox if party and constituency interests clash. The constituency she represents – say, an area that suffers disproportionately as a result of the financial crisis by way of mortgage defaults and unemployment – might be crying out for social intervention, while she and her party prioritise individualistic ideologies. She can recognise the validity of her party's insistence on loyalty while also acknowledging that people are demanding she acts in a way counter to her party affiliation. Pressure can be brought to bear by both sides. Party leaders can offer incentives or threats to maintain a particular approach by elected representatives, while constituents can threaten concerted and organised opposition at the next election. The relative claims of her party, her personal beliefs and her constituents must be weighed and prioritised. Consequently, during a period of economic expansion, said Congresswoman or Parliamentarian could push for minimal regulation of the mortgage market because it opens up the possibility of home ownership to people who have previously been excluded. The intention has a positive ethical dimension. However, when those homes are subsequently repossessed during a downturn she becomes the target of critics who say that she should have pressed for tighter rules surrounding mortgage and loan requirements. In one context (expanding economy) her actions are ethical; in a changing context (contracting economy) those same actions constitute her as unethical.

And fourth, there is the question of what political leaders are ultimately trying to achieve through their actions? For some it might simply be the practical outworking of their ideological beliefs. Others might want the satisfaction of seeing their particular interest group thrive, whether that group is the poor or the institutions of the financial sector. The religiously-motivated may declare that they have 'higher' intentions: eternity in heaven for an altruistic, faithful soul – a claim that stresses faith above proof and fact.[26] Yet others may have more venal interests: the trappings of power and the associated personal advancement. Almost every US president has, since the advent of Air Force One, mentioned in retirement that they miss that particular benefit of office. For all of the above – the altruistic, the selfish, the downright corrupt, the idealist – admitting to actions or inactions that might render them unelectable does not come easily in most liberal democratic political systems, if only because of the universal certainty among politicians that their opponent or replacement would make the situation worse.

In the field of national and global finance political leaders have been caught out by events often beyond their control, or at least beyond their interest to try and control, regardless of any post-event hindsight that might suggest otherwise. President George W. Bush was willing, motivated by party-political and ideological reasons, to advance extensive tax breaks to the richest stratum of American society. At the same time in the UK, Gordon Brown – first as Chancellor and then as Prime Minister – was willing to maintain a 'light-touch' approach to financial regulation because the banking sector was producing billions in tax receipts that could be redistributed to poorer sections of British society in line with his somewhat different political priorities.

Both approaches can be seen as greater or lesser social and moral goods or social and moral evils depending on the viewpoint of the observer and the priority given to particular truth claims. Asked why risk was allowed to escalate to the point where it precipitated the financial crash, the Governor of the Bank of England put it succinctly: 'It is very hard to say to someone who appears to be very successful, "What you are doing is potentially damaging to the rest of the economy".'[27] Such an attitude is made even more difficult to justify when, superficially at least, apparent social 'goods' are resulting from what is happening. In the US, millions of Americans who had previously been excluded from property ownership were included during the boom years. For some, the American dream seemed to come true, until it turned into a nightmare. In parallel, Gordon Brown expanded welfare spending in the UK faster than at any time in history, until the crisis made such expenditure unsustainable. In the financial crisis discourse, another group featured prominently – borrowers. And the remainder of this chapter will explore how they have been represented in the blame game.

When good borrowers turn bad

Former US Vice President Dan Quayle is attributed with saying that 'Bank failures are caused by depositors who don't deposit enough money to cover losses due to mismanagement.'[28] The accuracy or otherwise of the attribution is less important than the sentiment it contains: that the little guy – the apparently honest, law abiding citizen – has to share some of the blame with the dishonest, incompetent, opportunistic bankers who

lost the depositor's hard-earned dollars, pounds and euros. This was especially relevant when it came to the subprime mortgage element of the
financial crisis, as what started in the US spread rapidly around the world
with consequences that extended across the financial sector. A widely
accepted contributor to the crisis was the rise of house prices over several
years, which in turn encouraged both complacency and risk-taking in
the mortgage industry. Ben Bernanke pointed out in 2009:

> Lenders may have become careless because they, like many people at
> the time, expected that house prices would continue to rise – thereby
> allowing borrowers to build-up equity in their homes – and that
> credit would remain easily available, so that borrowers would be able
> to refinance if necessary. Regulators did not do enough to prevent
> poor lending, in part because many of the worst loans were made by
> firms subject to little or no federal regulation.[29]

Much has been said, and will be said for many years, about the role of
mortgage (and other) lenders in sanctioning loans that could not reasonably have been expected to be repaid. Countless reports, books and articles
explore the institutional behaviour at great length, including the failure of
regulators. In his dissenting statement accompanying *The Financial Crisis
Inquiry Report*, Wallison criticised predatory lenders and their unscrupulous
practices. However, he also went on to apportion at least some responsibility to 'predatory borrowers' who 'took advantage of low underwriting
standards' to get mortgages that they could not realistically afford to
repay.[30] The other dissenting Commission members also drew attention
to 'risky borrowers' who would not have been allowed similar loans in the
past. Elsewhere, Robert Schiller refers to 'complacent borrowers' who were
'naïve', unable to understand the complexity of the sophisticated financial commitments they were taking on, unable to appreciate that house
prices might stop rising.[31] The house price rises were a significant part of
the financial equations that would eventually fail, because the increase
in housing equity (which was assumed to be inevitable by lender and
borrower alike) would – it was almost universally accepted – make up for
any future repayment shortfall. History shows that it did not.

However, having briefly outlined the circumstances in which
borrowers, especially subprime borrowers, took out their loans, the
remainder of the section will consider how their identities, behaviours,
attitudes and ambitions came to be constituted in the crisis discourse,
as they were at least partly blamed for the circumstances in which they
found themselves.

In the 1980s in the UK, Prime Minister Margaret Thatcher promoted private housing ownership as part of her broader political agenda, introducing a policy that would eventually lead to 2.5 million publicly-owned council houses being bought by tenants and taken into the private sector.[32] In the US, from 1992 onwards, successive governments – Republican and Democrat alike – advanced a policy of encouraging private home ownership. In both countries, home ownership came to be seen – in a way that is still not the case in, say, Germany – as an inherent social 'good'. These policies and the ideologies that underpin them suggest that to own the house I live in, or at least to have a bank mortgage on the house I live in, is to somehow play a fuller part in society, to have a greater stake. Personal reasons for home ownership are as diverse as the individuals who make that huge financial commitment. For some it might be based on the pursuit of some kind of freedom – freedom from being at the mercy of a landlord who might evict at any time (within local legal constraints); freedom to do with my house as *I* want; freedom, eventually, from rent payments. Others may see home ownership as a physical manifestation of personal status and the fruits of ambition and success, not only the spectacular homes of the mega-rich but also the modest shelter in which individuals and families can take refuge – ideally safe refuge – from the world. The phrase, 'a man's home is his castle' (not to mention a woman's home being her castle – the original phrase invokes historical male-centric property laws) has graced the British and American lexicons for centuries.[33] Bricks and mortar thereby connoting concepts like permanence, rights, safety, and community membership in a way that transience and renting does not.

The political desire for expanded home ownership did not automatically have some magical effect on the sectors of society that had previously been excluded. In the UK, public housing was not simply offered at commercial rates to their tenants. After two years of tenancy, would-be home-owners became eligible for discounted prices, with discounts increasing for every year that the tenant had previously paid rent on a council (public) property. Long-term tenants could qualify for discounts of up to 60 per cent on houses and 70 per cent on apartments.[34] In the US, legislation through the Community Reinvestment Act was used in conjunction with the two government supported legal and financial entities, Fannie May and Freddie Mac, to extend credit to people who would previously have been seen as not creditworthy. In the 1990s and 2000s the private financial sector in the US loosened its lending criteria and encouraged mortgage-borrowing among the poor, partly because global trade imbalances had resulted in a glut of cash that needed to be 'put to work' for investors. Financially vulnerable sections of society were actively targeted by governments and financial institutions: individuals who took out subprime loans often did so because they had been persuaded it was a 'good thing to do'.

Subprime borrowers were not the only ones who contributed to the eventual crash. Individuals and families who already had mortgages or owned homes outright were encouraged to re-mortgage as a way of funding either more lavish lifestyles, pay down other debts elsewhere, or to afford the goods that consumer society presents as somehow deserved. In any case, so went the theory, rising house prices would counter-balance the additional loans and all would be well. Cake could be owned and eaten at the same time: the new rules of economics said so. Right up to the point at which millions found that they could neither eat the cake nor own it. Repossessions soared as the financial crisis hit. Devastated, confused, angry, betrayed, now homeless people were left wondering how it could all have happened – they only did what they were advised to do by politicians and financial experts.

Summary

The subprime borrowers, and others, did not wake up one morning and decide that they would play an active part in devastating the global economy. Most of them had devastated personal economies already. But they were offered a way out, a means of owning their own home – reassured by politicians and targeted by avaricious sales people in a bloated unconstrained financial sector. They became collateral damage in the financial truth wars as ideology, blind optimism, ignorance of history, faith in new and untested economic models, and inadequate regulation to which all major institutional actors and individual players turned a blind eye. Scott Stern, CEO of Lenders One, said to the Senate Banking Committee:

> Disclosures [of income by borrowers] were often less than adequate, and faced with a bewildering array of loan terms, borrowers tended to trust their mortgage banker or broker. The broken trust that resulted has damaged borrower confidence in the mortgage industry. I liken the situation to that of a doctor and patient dealing with a medical procedure. The patient bears some reasonable risk. But they don't bear the risk of malpractice by the doctor. In our industry, we have frankly seen too much mortgage malpractice.[35]

Stern's candour aside, what many are reluctant to admit is that, for several years at least, subprime borrowing was seen in many sectors of the political and financial landscape as an inherent good – poor people got houses, financiers made money, and politicians could feel a warm glow about the social equality they were achieving (and then get re-elected). Unfortunately, from Wall Street to Main Street, from Washington to London (and beyond), the codes which were supposed to prevent the crisis from happening were swept away when the tsunami of debt was eventually unleashed on the world. It is unlikely that any agreed version of events will ever materialise: political, financial, social and subjective truth claims cannot readily be weighed against one another. The personal and ideological perspective of the observer – every observer – will always skew the results of any analysis. If there is one financial truth that I suggest almost everyone can agree on it is this: the crisis was *definitely* someone else's fault.

8
Governing Greed

In 2005 profits at Goldman Sachs were reported at $5.6 billion, with pay and benefits for its staff members rising to $11.7 billion. The roughly 22,000 employees averaged $521,000 in total remuneration packages.[1] This was at one bank, in one year. Difficult though it is to recall through the fog of the subsequent global financial meltdown, bankers and other financiers were described variously at that time as gold-dust, inventive, dynamic, aggressive, buccaneers, risk-takers (in a positive, to-be-admired sense), demi-gods and many other now-ridiculous monikers. The *New York Times*, apparently without irony, referred to that year's annual Wall Street bonuses as determining who would take the title of 'Masters of the Universe'.[2]

Within a few years, reports of suicide among bankers would be greeted variously as sad, tragic, inevitable, or 'a good start', depending on the degree to which the speaker's life and livelihood was affected by the financial crisis. Similarly characteristic of the prevailing *zeitgeist* was a comment reportedly made by President Obama to a group of banking CEOs at the White House not long after he took office. In response to some of the unsatisfactory explanations being offered for the crisis, the President told his listeners: 'My administration...is the only thing between you and the pitchforks.'[3] In 2012, former British Prime Minister Tony Blair caused a fleeting re-eruption of anger – against him and against financiers – when he entered public debate once more and encouraged people not to think that life would improve 'if we hang 20 bankers at the end of the street.'[4] Strong though these sentiments are, they do not capture the full extent of the visceral hatred for bankers and other financiers that has been expressed across numerous countries and in many quarters of their respective societies. However, by the start of 2014 bankers and their bonuses were once more making the news as it was reported that the good times were returning – at least for them.[5] Further, in New York and

London bankers were hiring cinemas for private screenings of the film *The Wolf of Wall Street*. Away from prying eyes and critical voices they could lionise the excesses of 1990s financier and fraudster Jordan Belfort. Also at the start of 2014, it was announced that Goldman Sachs paid its staff \$12.61 billion[6] in 2013, a figure that looks startlingly like the 2005 figure above[7] – staggering sums to most observers.

After the initial implosion of economies around the world, politicians faced a dilemma over how to treat and speak about bankers. On the one hand bankers and other financiers became deeply distrusted and reviled, having caused huge political as well as economic damage. On the other hand, the bonuses they collect and the taxes they pay are highly attractive to finance ministers with deficits to reduce. The response of politicians in the US, UK and elsewhere has been to personalise the issue and emphasise the conduct and characters of individual bankers, adopting very direct language like 'spivs', 'gamblers', 'Trotskyite fantasies', 'unjustified bonuses' and 'social responsibilities'. Meanwhile, despite the highly emotive political rhetoric, American, British and other affected governments have been slow to enact tight laws and strict codes of practice, of the type threatened or promised at the height of the crisis, to ensure a similar banking collapse does not happen in the future.

Consequently, this chapter will analyse the personalisation of the politics of finance, examining how bankers have been portrayed, and have portrayed themselves, both before and during the financial crisis (I do not include 'after the crisis' as current levels of national debt in most major economies suggest that any short- term improvements should not yet be construed as post-crisis). Analysis of subjective aspects of banking will be addressed chronologically: first, the cavalier years between 2000 and 2006 where success seemed unending; second, the crash years – 2007 and 2008;[8] and third, the ongoing rebuilding phase, from 2008 until the present. The identities, conduct, and attitudes of senior bankers will be assessed against the codes that were supposed to constrain their excesses and with which they were expected to conform, as well as against the professional, ethical, social and other factors that feature in the financial truth wars that have waged over recent years.

The cavalier years in context

From 1980 onwards the US banking sector became increasingly deregulated; safeguards that had been in place for decades were gradually removed to encourage innovation and expansion in the industry.[9] After almost two decades of expansion – with the occasional contraction – the biggest change came about with the 1999 Gramm-Leach-Bliley Act (GLBA), otherwise known as the Financial Services Modernisation Act, which repealed the Glass-Steagall Act of 1933, a bill initially designed to encourage stability and confidence in Depression-era banking. The most important aspects of this change for banks and the bankers who ran them at the turn of the twenty-first century were twofold: it allowed them to merge with different kinds of financial institutions – securities or insurance companies for example, and it enabled them to borrow more to fund their activities. This in turn led to greater and greater profits, with remuneration packages to successful bankers following the same upwards trajectory. The extent of that trajectory would become synonymous with the excesses of bankers and banking CEOs over subsequent years.

The dangers were not only systemic but individualised. In 2006 Raghuram Rajan, Economic Counsellor and Director of Research at the IMF, argued that there was a 'behavioural' dimension to the risks that were being taken in the financial markets as fund managers and others sought to maximise profits at a time of low interest rates and high levels of liquidity (cash) looking for investment.[10] Put crudely, obviously even, Rajan highlighted a link between the way that incentives for investment managers were structured and the way that they subsequently behaved. With little return to be gained for investors from conventional investments like long-term bonds, increased risk-taking was inevitable if large sums of money were to be made by individuals and institutions. One of the ways of achieving greater returns was to borrow more – increase leverage – by taking debt out on debt: perhaps not *ad infinitum* but certainly to levels never seen before.

The crack cocaine of debt expansion was derivatives, whose purpose is not investment but a means of hedging (as in 'hedging your bets') against risk, which could range from natural disaster to shifts in interest rates, future prices of products, or rises or falls in inflation.[11] In the US, the Commodity Futures Modernization Act of 2000 (CFMA) deregulated the derivatives market and removed significant oversight regimes.[12] So from 2000 onwards this was to be an increasingly important area of global finance that reached into every part of every market, from mortgages to wheat crops to high technology. Ironically, the very stability that was promised by hedging turned out to be one of the great destabilising features of global finance within only a few years. This seeming paradox brought the writings of the late American economist Hyman Minsky back into vogue, especially his financial instability hypothesis: apparent stability would encourage complacency and risk-taking which, in turn, would inevitably lead to instability.[13]

Consider the rise in leverage levels after the turn of the century. At the end of 2000, against a gross market value of $3.2 trillion, the total value of outstanding derivatives on that market stood at $95.2 trillion. By 2008, just before the market crashed, outstanding derivatives stood at $672.6 trillion against a gross market value of $20.3 trillion.[14] An extended attempt to objectively summarise the systemic and institutional failures that brought the whole edifice crashing down are beyond the scope of this book. What remains, however, is an analysis of the ways in which the identity, behaviour and attitudes of bankers and other financiers were – and are – shaped and self-shaping: beginning with a brief look at how key banking figures reflected on that period when euphoria gave way to disaster.

Is sorry the hardest word?

As governments sought to understand how the financial crisis arose, CEOs were summoned to a number of inquiries: a precursor to taking steps that would prevent it from happening again. Goldman Sachs CEO Lloyd Blankfein was asked at a hearing of the Financial Crisis Inquiry Commission whether one of Goldman's pre-crash practices was 'proper, legal, or ethical'.[15] Goldman was accused of selling mortgage securities that they believed would default, then betting against their clients by 'shorting' those same securities – the fiscal equivalent of 'heads I win; tails you lose'. Blankfein answered, 'I do think that the behaviour is improper and we regret the result – the consequence [is] that people have lost money.'[16] Blankfein appeared to admit that Goldman Sachs had acted improperly, though he offered no apology and only expressed limited regret – regret that money had been lost, not regret for any personal or corporate actions that may have helped to cause the losses. Further, Goldman Sachs immediately put out a press statement 'clarifying' and denying any perceived admission of impropriety by its CEO.

The British parliamentary inquiry into the banking crisis also included personal appearances by the CEOs and chairmen of major UK banks that had to be rescued with public funds: Halifax Bank of Scotland and The Royal Bank of Scotland. In terms of blame and blame avoidance, it highlights different claims and counter-claims in the truth wars that broke out during the crash: bankers' enthusiasm to apologise for anything and everything except personal and institutional culpability; a concerted determination to shift blame elsewhere; and an eagerness to present themselves in the best possible light.

The opening exchanges of the inquiry were dominated by apologies, with the former bank leaders falling over themselves to appear contrite, while not actually admitting to anything untoward. Consider some of the words used and the sentiments expressed:[17]

Dennis Stevenson (former Chairman, HBOS)

We are profoundly and, I think I would say, unreservedly sorry at the turn of events. Our shareholders, all of us, have lost a great deal of money, including of course a great number of our colleagues, and we are very sorry for that...I would also say, Chairman, we are sorry at the effect it has had on the communities we serve.

Tom McKillop (former Chairman, RBS)

I am very happy to repeat [my apology] this morning. We were particularly concerned at the serious impact on shareholders, staff and, indeed, the anxiety it caused to customers.

Fred Goodwin (former CEO, RBS)

There is a profound and unqualified apology for all of the distress that has been caused.

Andy Hornby (former CEO, HBOS)

I am very sorry about what has happened at HBOS; it has affected shareholders, many of whom are colleagues; it has affected the communities in which we live and serve; it has clearly affected taxpayers.

Even in these opening words of the inquiry it is clear that no bank leader was willing to say, '*I* seriously misjudged the situation. *I* made crucial mistakes.' The apologies and regrets were targeted elsewhere. Owen Hargie, Karyn Stapleton and Dennis Tourish have studied apology avoidance on behalf of banking CEOs and highlight their inability or unwillingness to accept any degree of personal responsibility, their attempted identification with other 'victims' of the crisis, and limited expressions of regret.[18] So why would senior bankers not say sorry? Numerous motivations prompted the lack of contrition, or perceived lack of contrition, on the part of bank CEOs and other senior figures. These can be divided into three groups: legal, institutional and personal. The first of these is self-explanatory. CEOs would be very reluctant to say anything that might expose themselves or their institutions to legal action by investors who had lost money. Second, thousands of employees within the institutions would be implicated in any admission of culpability. Such an outcome would be bad for individual morale and institutional reputations: both of which are necessary for any ongoing organisation to retain if they want to continue in business. The third reason is in some ways the most interesting when personal and psychological factors come into play. After years of public lionisation, adulation, fear, envy and respect, the financial crisis relocated the 'Masters of the Universe' from their elite social stratum to a position most commonly occupied by vermin. Furthermore, any meaningful, fulsome apology would confirm that all the criticisms levelled at bankers and other financiers were therefore somehow 'true'.

When statements by senior bankers to the American Commission and to the Treasury Committee of the UK House of Commons are examined a number of contested truth claims come in to play. For example, Dennis Stevenson and Andy Hornby were presented with allegations from their former Head of Group Regulatory Risk, who claimed that he and his team had endured threatening behaviour while carrying out their duties. In response, Hornby shifted the blame elsewhere: 'I have absolutely no recognition of what he might be referring to. I assume you are referring to the period, way before I was chief executive.'[19] In summary: I know nothing; it wasn't me; maybe it was my predecessor. Criticism of HBOS from a 2005 review by the regulator, the Financial Services Authority (FSA), was then presented to the men who had been in charge at the time. The criticism said: 'There is a risk that the balance of experience among senior management could lead to a culture which is overly sales focused and gives inadequate priority to risk.'[20] Again, the criticism – and culpability – was explained away:

> We went through everything with [the FSA] in huge detail...we had colossal banking experience...we did try and make radical changes. We have all accepted the fact that the balance sheet growth that had been built up over many, many years meant that we did end up over-reliant on wholesale funding. We would have liked to have predicted an even greater contraction in wholesale funding markets. I think it would be difficult. I accept the fact we did not fully prophesize it. I do not believe that was because we were not listening to the risk function.[21]

The question of why the banking executives behaved in the way they did was stated very transparently by McKillop and is representative of attitudes on Wall Street, the City of London and beyond: institutional shareholders wanted the company, and profits, to grow. 'I would say the drift from the most institutional shareholders was increase the dividend, share buybacks, return capital, do not sit on capital and run a very efficient balance sheet.'[22] Expanding leverage was the way to achieve all of the above – right until the edifice came tumbling down.

Shaping bankers' behaviour

For more than two decades before the crash a series of decisions by governments, regulators and central banks helped to create the conditions in which the worst excesses of bankers' behaviour could manifest itself. Apart from the steady stream of financial deregulation, some of the most behaviour-altering influences came from specific interventions by the US Federal Reserve (the 'Fed'), the Bank of England and other central banks during times of financial crisis – thereby helping to shape bankers' attitudes to risk. When the stock market and its associated futures and options markets crashed in 1987 with the accompanying danger of widespread recession (or even depression if things were allowed to spiral out of control), the speed of events prompted the Fed to take decisive, high profile action to stop contagion spreading. Key activities included: injecting liquidity (cashflow) into the market; improving short-term credit conditions; boosting market, public and institutional confidence by openly stating its ongoing engagement with the crisis; encouraging well-found banks to extend support to other sectors of the financial markets.[23]

A decade later, a series of events in emerging Asian markets, falling oil prices, a lack of robust internal financial mechanisms and poor international credibility caused a crisis in Russia. In August 1998 the Russian government defaulted on its domestic and foreign debts.[24] With the IMF refusing to bail out the Russian economy, uncertainty spread, from emerging economies to the US equity markets, and the dollar fell in value. The Fed and other central banks responded by lowering interest rates, thereby lowering borrowing costs and stimulating demand for goods.[25]

The biggest financial upheaval of that era came in the aftermath of 9/11. Initially, as a result of a combination of national panic, partial failure of payment processing systems,[26] and a desire to stop the physical and psychological damage of the terrorist attacks spreading to the markets, the New York Stock Exchange (NYSE) was closed for a week. Markets everywhere became jittery and when the NYSE was opened again for trading, stocks fell. Consumer confidence and consumption also both fell.[27] To strengthen confidence the Fed once more acted decisively. It increased liquidity in the system to ensure that payments could be made, bank lending was stimulated, and investments in the economy would not be disrupted.

The macroeconomic detail of these events is beyond the scope of this discussion. However, there are two crucial elements of the interventions by the Fed and its counterparts that are important in understanding the extent to which bankers were affected by those events and that shaped their behaviour – negatively – as a result. One of these effects was intentional while the second was not. Primarily, the central banks in general and the Fed in particular were keen to iron out these shocks to the financial system to prevent any medium-to-long-term downward spiral. While each central bank was concerned with its national interests, the globalisation of finance and the interconnectedness of markets meant that they had an international impact as well. The secondary effect of these interventions was not on the banking *system* but on the people who run it, manipulate it and generally make money out of it: bankers. Either implicitly or explicitly, they came to understand that in the event of a major crisis in the system, the central banks would become lenders of last resort and stop the whole financial edifice from collapsing. The dominant financial truth was, and remains, an unlimited commitment by political leaders to ensure the financial system – warts and all – does not disintegrate.

On its own, the willingness of the Fed to intervene in the markets repeatedly indicated to those who ran the banks and worked in them that they could assume a degree of risk in the system was, *in extremis*, already being offset. When that element of risk offset was set alongside the confidence (which we now know to have been misplaced) prompted by hedging against risk in the derivatives markets, a significant element of the framework for catastrophe had been put in place. Consider some of the other elements in play: shareholders were demanding higher and higher profits from the banks; a global glut of cash that meant greater risk had to be borne to keep profits up; new financial models and algorithms apparently negated that risk anyway; and higher leverage became a way of life. Amidst all of the foregoing, possibly the most significant reassurance to be found in the whole financial system, the one that shaped institutional policies more than any other, is this: the lender of last resort will step in if things go seriously, seriously wrong. Low-level risk was covered; medium-level risk was covered; and exceptional-level event risk was covered.

Of the many other major and minor contributions to the financial crash, one had more direct bearing on the conduct of individual bankers than any other: the way that they were, and in many cases continue to be, paid. The cornerstone of the pre-crash compensation culture for bankers was the annual bonus. Bonuses were calculated on a series of short-term results and paid out largely in cash at the end of the bonus year. When traders made high risk bets with money belonging to investors and bank depositors, if the bet made huge returns so did they. However, when high risk wagers failed, there was no punitive financial counter-measure levied on the trader – basic salary would still be paid. Risky behaviour reached its peak as the tsunami of over-leveraged institutional debt was about to break over the markets in the US and elsewhere in September 2008. In one of the most contentious cases, the US Financial Crisis Inquiry Commission identified the key contributory factors to its failure: poor governance, inadequate risk management, and a compensation culture that focused on short-term profits.[28] Meanwhile, in the UK a similar conclusion was being drawn by the Financial Services Authority (FSA). Adair Turner, the FSA Chairman, reported:

> There is a strong prima facie case that inappropriate incentive structures played a role in encouraging behaviour which contributed to the financial crisis…past remuneration policies, acting in combination with capital requirements and accounting rules, have created incentives for some executives and traders to take excessive risks and have resulted in large payments in reward for activities which seemed profit making at the time but subsequently proved harmful to the institution, and in some cases to the entire system.[29]

Bankers' bonuses were only part of a much larger and complex financial equation, the details of which will never be universally agreed because of the political, commercial, ideological and subjective elements that shape the crisis discourse. However, they definitely shaped the behaviour of many individuals whose combined actions were highly significant. The policies of governments, regulators, central banks, shareholders and individual institutions combined not to *restrict* the excesses of individual traders and other banking executives but to *encourage* it. In that environment, financially-motivated bankers would have been behaving irrationally if they had not put their snouts in the trough and gorged themselves. The uncomfortable question is this: who wouldn't have done the same? President Obama summed up the situation in 2009:

> We have lived through an era where too often, short-term gains were prized over long-term prosperity; where we failed to look beyond the next payment, the next quarter, or the next election. A surplus became an excuse to transfer wealth to the wealthy instead of an opportunity to invest in our future. Regulations were gutted for the sake of a quick profit at the expense of a healthy market. People bought homes they knew they couldn't afford from banks and lenders who pushed those bad loans anyway.[30]

The list of culpable individuals, institutions, policies and regulations (or lack thereof) is extensive. However, if the discussion of events so far has in any way implied that the significant changes which took place in the lead-up to the financial crisis somehow 'just happened' – deregulation, changes in law, huge growth in leverage levels, innovative financial products that were not understood by their users – the remainder of the section will show how these were not some sequence of unfortunate, unrelated incidents.

Each of the legal and institutional changes that took place since the 1980s were planned, lobbied for, and implemented by bank and other financial sector leaders at every stage. The speed with which changes occurred was related to the party in office at the time, the extent to which the President or Prime Minister was motivated by the ideology of *laissez-faire* capitalism and wider social and economic conditions, and tolerant of the corporate aggression of the institutions involved. The US National Commission summed up the collective activities of top-level bankers and financiers as follows:

> The financial sector, which grew enormously in the years leading up to the financial crisis, wielded great political power to weaken institutional supervision and market regulation of both the shadow banking system[31] and the traditional banking system. This deregulation made the financial system especially vulnerable to the financial crisis and exacerbated its effects.[32]

At this point in the discussion – and in the formal investigations into what went wrong – the financial truth wars become more guerrilla warfare than full-frontal assault. The very nature of political lobbying is secretive, deniable and even less regulated than the banking activities it helped to liberalise to the point of self-destruction. What I am willing to say with maximum confidence and minimum evidence – for obvious reasons – is that at every stage in the deregulation that took place on the road to ruin, major banks and other institutions were using their financial power and political access to lobby for the changes in law that took place. Further, when the changes on offer were for great or greater liberalisation, lobbyists for the financial sector pressed for the latter.

In 2000 the Federal Reserve examined the rules that would protect borrowers from predatory conduct by financial institutions. A tension existed however. The choice was between strict rules that would definitely protect individuals and institutions but which would also limit growth in the financial sector, and looser rules that could be navigated more easily by the unscrupulous. Alan Greenspan advocated the latter approach, reinforced by the application of existing laws on fraud and other illegal activity. He wrote in evidence:

> Low-down-payment loans to borrowers with limited savings but adequate income to support the monthly payments might be perfectly appropriate, while the same loans to borrowers who cannot document their income may not be. In short, these and other kinds of loan products, when made to borrowers meeting appropriate underwriting standards, should not necessarily be regarded as improper, and on the contrary facilitated the national policy of making home-ownership more broadly available.[33]

It was deliberate social policy – underlined by political ideology that prioritised market self-governance and commercial expansion over stability and imposed constraint – not to write tight protective rules. Institutions would be given freedom to interpret much looser guidelines when deciding how to target borrowers and sell them mortgages and other products. This decision even had what appears to be a positive moral dimension: supporting the national policy of extending home ownership. Subsequently, between 2000 and 2006 – the height of the boom years and the financial free-for-all – the Federal Reserve prompted legal actions against only three institutions for improper mortgage lending: two American banks and one French bank.[34] This reluctance to create or enforce strict codes of practice not only shaped the legal position of institutions but the ethical conduct of individual bankers.

Rules are not enough – the ethical dimension

One year after the crisis was at its 2008 peak and Lehman Brothers bank collapsed in the biggest bankruptcy in American history, President Obama went to Wall Street to outline his vision for how the finance industry should work in the future. He told the gathered ranks of bankers and others:

> Those on Wall Street cannot resume taking risks without regard for consequences, and expect that next time, American taxpayers will be there to break their fall. And that's why we need strong rules of the road to guard against the kind of systemic risks that we've seen. And we have a responsibility to write and enforce these rules to protect consumers of financial products, to protect taxpayers, and to protect our economy as a whole ... these rules must be developed in a way that doesn't stifle innovation and enterprise.[35]

President Obama's rhetoric is consistently eloquent and, like Clinton and Blair before him, makes forceful points where necessary while gliding effortlessly over the contradictions. Obama's key point in this excerpt, and in other parts of the speech where he refers to the previous 'lack of clear rules' and the 'gaps in the rules', is that new rules and financial regulations must be created to ensure that the financial crisis cannot be repeated. The excessive behaviour of the bankers and others who caused the collapse in the first place must be curtailed. Protection would thereby be extended to consumers, taxpayers and the economy. However, the contradiction soon follows: innovation and enterprise are not to be held back. The whole purpose of rules and regulations is to stifle particular, extreme behaviours, to ensure that innovation is not allowed to roam freely and unconstrained as it had been previously. And even if new rules can be agreed – despite intense lobbying from the banking sector to continue to advance financial liberalism – there is a philosophical problem with seeing them as a solution to the problems that went before: rules on their own cannot be relied upon to prompt ethical conduct.

There are two dimensions to morality as it has been used in this book – the first is code-oriented, while the second concerns personal ethics. Code-oriented morality is about conforming to rules, regulations, laws, and so on. Personal ethics is about individual choices and the beliefs or philosophical assumptions that underpin them, the individual and social motivations for behaving in particular ways, and the ultimate ends (heaven, a good life, social conscience) that motivate each person. Therefore, it is an individual ethical choice whether or not to conform to particular rules or codes, such as driving laws or financial regulations. Each person chooses whether or not to conform to the speed limit, and then for a multitude of reasons – only one of which is the danger of being caught breaking the rules and then hit with a suitable punishment. National leaders like Obama, Cameron, Merkel and others, proposing new rules and laws is one thing; passing those laws in the face of concerted pressure from banking CEOs who want to maintain maximum freedom of action with minimum consequences and costs is another. Then, maintaining those laws against lobbyists and reinforcing those laws in their respective judicial systems will be the next challenge. Ultimately, it is not possible to govern greed – merely to ameliorate its worst effects.

Away from the code-oriented, rule-making response to the financial crisis Christopher Bones suggests that corporations who want to be successful in the long-term – and banks offer a paradigmatic example – need to put people in positions of responsibility who know the difference between right and wrong.[36] Eloquently simple in theory but difficult to achieve in practice when in relation to finance (and every other sphere of life) right and wrong are as contested as the philosophies and truth claims that underpin them.

Consider the question: What is the ultimate aim of a trader in a major bank? There are not too many mainstream religions whose doctrines encourage adherents to make the maximum amount of money possible in the shortest space of time, ignoring social consequences in the process (though some Christian denominations do promote a 'health and wealth' gospel that believes riches are an indication of God's blessing). High risk securities' trading is probably not regarded as a short-cut to heaven by those who do it. The motivations for those involved are much more likely to be materialistic and based on earthly rewards in this life rather than spiritual rewards in the next. However, if someone is primarily motivated by money, new regulations and constraints on financial practices might place limits on that person's behaviour (if enforced) but they will not fundamentally change the personal motivations at work. Any easing of regulations or failure to enforce them will inevitably lead back to previous (mal) practices or new exotic variations. In addition, pressure from shareholders to maximise returns on investment will not be reduced. Extending complicity further, individuals who belong to private pension schemes will expect their fund managers to maximise profits so that pension payouts will similarly be optimised in due course. Behind all of these stands the Inland Revenue or the IRS who wish to exact maximum taxation, directed in turn by governments who have ideologically-driven political ambitions for the monies involved. The commodities trader might be the person at the front end who is feeling the pressure to produce profits, but that pressure is built up by a vast web of interrelated interests, and self-interests, that span institutional, societal and political expectations. Perhaps countries simply get the bankers they deserve.

The next question I would ask is: Why then should bankers conduct themselves ethically? Two related calculations which they will have to make concern risk *versus* reward, and benefit *versus* punishment. If regulatory measures are put in place to reduce the risks that led to the financial crisis then it would seem reasonable that bankers should subsequently expect lower rewards. However, post-crash behaviour at surviving institutions suggests this is not the case. In 2010, David Cameron told a European Union summit: 'Bankers have to realise that the British public helped to bail out the banks and it is very galling when they see bankers pay themselves unjustified bonuses ... the banks have got to think about their social responsibilities and their wider responsibilities when they make these decisions.[37] Similarly, a year after the crash President Obama spoke of 'restoring a willingness to take responsibility'[38] in the banking sector. The 'reward' part of the bankers' calculations does not appear to have fallen as far as the expectations of the public or the expectations of political leaders towards banker remuneration.

The coercive weapon in the regulatory armoury is punishment. Much has been said on this matter. President Obama warned: 'We can't allow financial institutions, including those that take your deposits, to take risks that threaten the whole economy.[39] Even more forcefully, David Cameron stated: 'I have always said with respect to bankers that if they get it wrong, they should, like anybody else, feel the full force of the law.'[40] Tellingly, however, according to the UK's Financial Conduct Authority, 5,873 'rogue' managers and staff were ejected from financial institutions between 2008 and 2014.[41] Yet no British financier has been jailed for their part in the collapse. Fred Goodwin was given a knighthood in 2004 for 'services to banking' before he led the Royal Bank of Scotland into financial meltdown and nationalisation in 2008. Stripping him of his knighthood appears to be the high watermark of British establishment retribution against bankers. It is hard to impose punishments that 'fit the crime' if no rules were broken – because the rules were so lax and easy to navigate for creative, highly motivated bankers – and therefore no crime committed.

Following the crisis and in response to the pleadings and not-so-veiled threats from political leaders around the world, reaction from financiers has been consistent. Simon Lewis, Chief Executive of the Association for Financial Markets in Europe, responded to proposed changes to Markets in Financial Instruments Directive (MiFID), which regulates the financial services sector across Europe:

> Regulation needs to keep pace with changing market practices and technological developments, so it is right that existing rules should be reviewed and we agree that regulatory improvements are needed in some areas. But it is vital that any proposed solutions address real problems, are supported by robust impact analysis and maintain user choice, innovation or competition. Getting this balance right will be a difficult task for the regulators, but the financial services industry, both 'buy side' and 'sell side', is committed to helping them do so.[42]

Lewis acknowledges that changes in market practices and technological advancements are forging ahead of regulation and he concedes that regulators need to keep up with those changes. There is no suggestion that such innovation should be subject to regulation before it is allowed to proceed in the first place, though he concedes that 'regulatory improvements are needed in some areas'. These areas are unspecified but constitute only 'real' problems, presumably defined as such by the finance industry and contrasting with trivial problems or unnecessary interference from the authorities. All of which indicates a superficial willingness on the part of the financial sector to be suitably governed.

The real battleground lies in the suggestion that any solutions should be 'supported by robust impact analysis and maintain user choice, innovation or competition.'[43] Any such analysis cannot be conducted with the objectivity claimed by either the financial services sector or the politicians who would set the terms of future regulation. Facts, trends, analysis and responses will all be interpreted through the lenses of economic ideology, fiscal policy, and institutional and personal interests. It is significant that President Obama also called for rules to be 'developed in a way that doesn't stifle innovation and enterprise.'[44] If there is a truth war being waged between politicians and financiers then it is increasingly becoming a phoney war – even the language they use sounds the same, and in some instances *is* the same. Political leaders and bankers are united in their response to the crisis: they are all doing everything possible to sound like they are doing everything possible to ensure it does not happen again. Meanwhile, as Nouriel Roubini – the original Dr Doom – and Stephen Mihn point out, lawmakers appear to have grown comfortable with the *status quo* whereby big bonuses are back, minimal reforms have been put in place, and significant structural weaknesses remain: as does the towering debt that resulted from the crash and the global bailout.[45]

When leadership of the Federal Reserve moved from Ben Bernanke to Janet Yellen in early 2014, numerous warnings were being made in the US. As he left office Bernanke spoke of achievements like the Fed's rapid and forceful response to the liquidity shortage in the system as the crisis hit; bank debt guarantees made by the Federal Deposit Insurance Corporation; and the injection of public monies into banks and other financial institutions.[46] However, he also warned of tasks that had yet to be completed: full implementation of new regulations; strengthening of oversight regimes; and domestic and international coordination of efforts to deal with future institutional failures. Yellen's first comments in office in March 2014 caused downward movement on the markets when she merely suggested that the Fed would press ahead with reducing the amount of liquidity it was injecting into the financial system. At the same time in Europe, Mark Carney, Governor of the Bank of England, pointed out that one of the key drivers of the credit boom in the 2000s was to a significant extent still in place: a plentiful (if reduced) supply of cash available at low cost. He warned: 'the period of low and predictable interest rates before the financial crisis helped drive a "search for yield" and leverage cycle, even with inflation subdued. It doesn't take a genius to see that similar risks exist today.' [47] None of these responses – *six years after the crash* – suggest that sufficient regulations are in place to reign in the behaviour of bankers whose expectations appear to be little changed in the aftermath of the worst financial catastrophe since the Great Depression.

Summary

There are, of course, sound arguments for the length of time taken to achieve the safeguards that will more fully protect the global economy from a repeat of events of the 2000s. These include: an initial focus on stemming the crisis; helping institutions survive; trying to recover consumer and market confidence; not wanting to lose too much of the 'growth' that had been achieved previously; and seeking to adequately understand what happened. However, there are also profound reasons why the promised/threatened level of financial regulation has not been implemented: ideologically-charged political inertia on Capitol Hill, in the City of London and elsewhere; the risk of 'tightening up' in one country, leaving cash and expertise to flow to another less regulated location; the political leverage of a banking industry with short memories, deep pockets and an unreconstructed appetite for high profits; and individual financiers whose personal motivations are little changed (excessive greed, as it continues to be called in the popular press). Some countries are under greater pressure than others to grant banks maximum freedom and increase the associated risk they carry. The Basel III agreement on the regulation of capital sets specific and minimum requirements for institutions in terms of liquidity levels and leverage ratios. It also requires governments and central banks to set and observe 'macroprudential policy'[48] to prevent failures in the system becoming amplified through financial contagion or human panic.

These new guidelines may be an improvement over what went before, but they cannot change human nature and the desire to acquire more and more wealth. They also cannot change the underpinning assumptions through which individuals, institutions and societies view the accumulation of wealth and shape related policy. Personal ambition, the pursuit of power, narcissism, predisposition to risk-taking and – on the margins of behaviour – naked criminality continue to shape conduct and attitudes. In addition, politicians will never be able to regulate the financial industry stringently enough to satisfy those who have suffered the most from the collapse. Conversely, they will not be able to regulate it lightly enough to satisfy the bankers who will always want to embrace freedom of manoeuvre while avoiding personal responsibility for losses. The result is that political leaders can never be entirely truthful with either party – the public or the finance industry – and will always try to opt for a course of action that sounds, again, very much like having cake and eating it while not truly being able to afford to own the cake or enjoy consuming it. The financial truth wars will therefore continue unabated because ultimately there is no foolproof way to govern greed.

9
Who Mentioned the War?

In 2010, as the Greek economy wavered on the edge of a financial precipice and the scale of its financial crisis could no longer be hidden from the public by policy makers and bankers, an internet joke went viral. It went something like this.

German Chancellor Angela Merkel arrives at Athens airport on her way to meet with the Greek Prime Minister to discuss a European Union bailout of the collapsing Greek financial sector.

Customs Officer to Chancellor Merkel: 'Occupation?'
Chancellor Merkel: 'No, just a business trip.'

The words, of course, allude to the German occupation of Greece in World War II, and in those few seconds ancient enmities were given a new articulation in the twenty-first century. Cultural stereotyping that had, for years, been the source of humour between Europeans – almost in the way that families have private jokes that outsiders do not understand – began to take on a darker, more menacing edge. The Merkel joke was subsequently given Spanish and Italian variants, with the Cypriots offering the most bellicose and anger-fuelled version as they queued outside their banks in 2013. These resorts to old attitudes were not confined to the Mediterranean fringe of Europe. In Germany, industrious citizens (another stereotype) began to ask why they should be working overtime to pay for Spaniards and others to enjoy an afternoon siesta.

This chapter examines some of the possible ways in which Europeans might form or re-form personal identity, behaviour and sense of purpose in light of the financial crisis and the reordered world in which they find themselves, and in a Europe where difference is no longer a familial in-joke but an increasingly harsh dimension of European politics. Mutual antipathy has expanded at the expense of mutual regard and trust. Identity is no longer formed solely or straightforwardly (perhaps it never was) in relations between self and other, where the other is located in neighbouring states or in the institutions of the European Union. The other is also found in the relationship with oneself: in that gap between who 'I' am today in relation to the 'I' whose place in the pre-crash world is no longer recognisable.

The first section will begin with an outline of the philosophical challenge of trying to say something meaningful about a subject matter in which I as the author am personally implicated as a British and European citizen. The remainder of the section loosely traces the development of key institutions from their fragile roots in a devastated post-war Europe to the political dominance they wield over hundreds of millions of citizens in the twenty-first century. European politics is considered as a domain in which identity is formed and re-formed over time: drawing attention to the shifting power dynamics and competing truth claims between the centralising tendencies of the institutions of the European Union (EU) and the sovereign interests of its member states and their individual citizens. The second section will argue that since World War II, the EU has increasingly transcended mere institutions and legal processes to the extent that its supporters see it as a form of moral order whose existence and purpose is self-evident and whose pronouncements dominate the financial truth wars and to the point where questioning that moral order and its supporting ideological and institutional truths is more akin to religious heresy than political debate, leading to a resurgence of the language of good and evil within Europe. The final section explores how historical stereotypes are being resurrected in the formation of personal ethics and identity as the actions of EU decision makers provoke the kind of antagonisms and tensions that the institution was originally created to prevent.

Unity and disunity in Europe

In an interview in 1983 Foucault rejected the notion of a fixed subject whose identity, values and behaviour could be settled and summarised. He set out to show how a person's character and identity could be shaped – and self-shaped – in specific ways: as mad or healthy, delinquent or non-delinquent.[1] The setting for two of Foucault's studies were the asylum and the prison as he explored how particular truth claims were used in conjunction with practices of power and 'games of truth'[2] to shape individual conduct and a sense of self.[3] This section will examine how an individual comes to be seen as ethical, and can form himself as ethical, in the unstable rivalries that characterise competing truth claims and practices of power within the European Union: specifically, in relation to the idea that the EU is or has a moral order that is inherently superior to any alternative political possibility.

In setting out an overview of some of the power dynamics at work in the European Union, there is a degree of ambiguity in the presumptions with which I commence. Philosophically, I reject the notion that I speak definitively from some objective vantage point where I can look down on all the European worker ants and assess their vast, joint endeavour. However, I cannot write from 'Nowhere', as it were, similarly untouched by the world. I will therefore try to be as honest as I can in acknowledging that I write from within the very political framework I am critiquing. Further, I write from a British perspective at a time when the UK is experiencing considerable national tensions on the European issue: it is part of Europe yet, being an island, somehow not part of Europe. Popular opinion increasingly demands a say – a referendum – on the UK's relationship with the EU, while the mainstream political parties have tacitly agreed for 40 years to avoid such an eventuality. So I will attempt to interpret social, political and ethical meaning in a changing political environment that often prefers comfortable, artificial certainties to the uncomfortable realities that keep appearing. This might all sound a bit abstract, a bit airy-fairy, to individuals who prefer a life of certainty: *I know what I believe and I believe what I know.* However, such *faux* certainty has no place where an unforeseen financial tsunami wreaked havoc on not only the financial institutions of Europe but on the way individuals see themselves, their country and the place of both in the wider world.

For many realists – both in the sense of political realists and in the more generic characterisation of someone who is sensible and practical – the kind of instability of meaning that I acknowledge and try to operate within is sufficient, in and of itself, to preclude its legitimacy as a form of knowledge or at least attempted understanding. I cannot in any way, even without the word constraints of a chapter such as this, 'prove' such a critic wrong. Paradigmatic choices are made and intellectual battle-lines defended as surely as the trenches that defined battle in World War I. As this chapter proceeds, however, I can point to the instability and contingency of the political structures under consideration and the consequent subjectivities – identities, values, behaviours – that reflect those wider political uncertainties. Any attempt to bring what looks like certainty into a domain of political and economic uncertainty is at best a futile endeavour and at worst a fool's errand.

The second key presumption with which I commence is that while the antecedents of the European Union (EU) came into existence for the higher purpose of bringing together European states as a means of avoiding a repeat of the 1930s and the rise of Nazism or other extremism, over time the expansion and increasing federalisation of the EU has become an end in itself. The view from the European centre has increasingly been that its existence, growth, extension of powers, and accumulation of means by which to control member states – most notably through the European Court of Justice, the European Court of Human Rights and more recently the European Commission – is inherently good. Federico Mancini describes how the jurisprudence of the European Court of Justice '*read into* the Treaty [of Rome, 1957] an open-ended charter of fundamental rights', highlighting how the Court has taken unto itself decisions and powers that were previously the domain of national sovereignty.[4] Further, the Treaty of Rome opens with words that are unambiguous in their intention, with its signatories 'Determined to lay the foundations of an ever-closer union among the peoples of Europe.'[5]

Mancini goes on to highlight one aspect of ever-closer union: that in recent years the powers of EU member states have been eroded. He points to 'the election of the European Parliament by universal suffrage' as a step towards reducing the democratic deficit in Europe, even though it retains only a small proportion of the EU's lawmaking capability.[6] In addition, the need for elections to the European Parliament was 'prompted by the awareness that the growing range of the Community's[7] powers had significantly detracted from the sovereignty of national Parliaments.'[8] Over time, more and more powers have been withdrawn or ceded – without consultation with individual citizens of European states at the ballot box – from national parliaments in a way that demonstrates a shifting balance of power from member states to the institutions of the EU itself. Lest any non-European gets the impression that the combination of the words 'universal suffrage', 'election', and 'European Parliament' somehow suggests a fully democratic process at work, that misunderstanding should be disavowed. Elections to a body that wields minimal power in a much larger edifice is akin to allowing gladiators of old to elect who gets to go into the Colosseum first to fight the lions. A vote may have occurred but the lions are still hungry, the baying crowd is demanding blood, the word of the Emperor is absolute, and nobody is coming out in one piece. The power relations remain undisturbed. In May 2013 the British Foreign Secretary was scathing in his indictment of the democratic deficit in Europe when he declared: 'if the European Parliament were the answer to the question of democratic legitimacy we wouldn't still be asking it.'[9]

This clawing of power to the European centre was neither random nor unintentional. As early as 1948 the European Union of Federalists gathered at the Congress of Europe at The Hague with the ambition of authoring, and then seeing enacted, a European federalist constitution that would create a bulwark against any future rise of nationalism of the kind witnessed from the 1930s until 1945. Events after World War I sowed the seeds of the horrors that would come to fruition in World War II and there was concern across Europe that the pattern should not be repeated. The moral intention in such an undertaking was and remains highly convincing, assuming that the moral argument was not being used as a means of inciting citizens to support a course of action whose primary intention was political. Given the overlapping and competing moral and political interests, there were disagreements about how a peaceful Europe could be achieved. The UK led the rejection of a federalist constitution at that time, preferring a less structured relationship between European states that could still provide the desired level of political stability. What emerged was the newly-formed Council of Europe (a coalition of nations), which enacted a binding Charter of Fundamental Human Rights that represented the first step towards federalism.

The federalist trajectory in European politics was under way. By 1963, for example, French President Charles de Gaulle vetoed the UK's application for membership of the European Union's predecessor, the European Economic Community, on the basis that the UK was insufficiently committed to European integration.[10] Critics of the UK and other reluctant federalists might observe the prescience in de Gaulle's comments. For over half a century this federalist approach has dominated European politics, underpinned by a fear of the return of the evil of Nazism or the emergence of some new ideology that could cause social division and extreme otherness. Viewed in opposition to such a possible eventuality, it is easy to see why the EU's federalists see ever-closer political union as an inherent and unquestionable good: initially aimed at the annihilation of extreme difference, and subsequently extended to the elimination of difference generally, all in pursuit of a common European identity.

However, the success of the European project – where success is defined as the centralised accumulation of power and increasingly subordinate member states – has led to institutional egotism and consequent claims to moral order that are presented in self-referential terms: something is good if it is *of* the EU, or if it is commanded *by* the EU. By implication, any opposition to the EU as it currently stands or opposition to the goal of further European integration is represented as foolish, immoral or worse. The inherent 'good'-ness of the EU has long stopped being argued for by its proponents; it is now an *a priori* assumption that underpins all deliberations and all actions. On the rare occasion where a member state has held a referendum and the 'wrong' outcome has resulted (that is, an outcome that challenges the EU's institutional ambitions), that state has been invited to hold a further referendum to secure the 'correct' result. Ireland's referendum on the Nice Treaty in 2001 – which would delegate further powers to the European Union – produced an unexpected rejection by the usually pro-European Irish voters. However, a re-run of the referendum in 2002 ratified the Nice Treaty in a manner that was popularly (from a pro-EU perspective) described as producing the 'right' outcome. Subsequently in Ireland, the 2008 referendum on the Lisbon Treaty provided an even more decisive rejection of the trend towards centralisation of power in Europe. Yet by the end of 2008 it became clear that a 'Lisbon II' referendum would take place the following year. Jane O'Mahoney says of these events:

> In the Nice I and Lisbon I referendums ... a number of effectively organized groups and political parties on the margins of the political system, often espousing populist and anti-establishment ideas, succeeded in capturing the referendum agenda. These actors capitalized on the fears and distrust of an electorate deficient in general knowledge about the EU and were facilitated by the complexity of the issues at stake and the institutional rules of the Irish referendum game.[11]

O'Mahoney places significant emphasis on the voters' degree of knowledge about the EU, exacerbated by the complexity of the key issues facing them. Her assumption appears to be that greater understanding by voters of the EU would tip the balance in favour of further centralisation. In making such an argument O'Mahoney interprets events in a way that is common amongst those who would pursue the ideological goal of complete political, social and economic integration. She appears not to give due weight to the possibility that the political direction of travel towards federalism is actually understood perfectly well by voters who have decided that they do not want to relinquish any more state sovereignty to the EU and its institutions. As a result, Irish voters rejected the inevitable shift in the power relations between Ireland and the EU from the national government to Europe. Twice. O'Mahoney implies that such an outcome is undesirable in itself, using the terms 'populist' and 'anti-establishment' in a derogatory and negative manner. This is ironic given that the winning side in any and every referendum, or other truly democratic vote, must be the most popular with those who cast a vote: hence 'popular', which shares the same etymological root as 'populist'. More likely, and I will risk a generalisation here, the term 'populist' is used when the proposer of a particular motion or course of action wishes to denigrate an opponent who does not see or agree with the assumed innate superiority of their position.

British voters have, for decades, not been trusted by successive governments with a referendum on the relinquishing of state sovereignty to the EU because the electorate might not produce the 'correct' – that is, pro-EU – outcome. Hence the 2013 statement by William Hague, British Foreign Secretary, that 'Too often, the British people feel that Europe is something that happens to them, not something they have enough of a say over... Trust in the institutions is at an all time low. The EU is facing a crisis of legitimacy. I do not believe that we are going to solve this through more powers for the European Parliament.'[12] Hague's comments merely echo Prime Minister David Cameron's statement a few weeks earlier in January 2013 when he said: 'there is a growing frustration that the EU is seen as something that is done to people [across Europe] rather than acting on their behalf. And this is being intensified by the very solutions required to resolve the economic problems.'[13] Cameron identified increasing frustration not just in the margins of the economic strugglers like Greece, Cyprus and Spain but in the economic powerhouses as well: 'We are starting to see this in the demonstrations on the streets of Athens, Madrid and Rome. We are seeing it in the parliaments of Berlin, Helsinki and the Hague.'[14]

Note, however, the way in which Cameron referred to the individuals and countries concerned: in Athens, Madrid and Rome it was anxious and angry *citizens* who took to the streets in popular protest against what they saw as the EU's self-serving and anti-democratic behaviour. Meanwhile in northern Europe it was the *parliaments* that were most concerned: the German Parliament, the Finnish Parliament and the European Parliament. The concerns, however, were not all the same. The German and Finnish parliaments were worried about their increasing financial liabilities to southern European states, while the European Parliament was worried about previously unseen levels of popular hostility to the EU.

It was not only rioters on the streets in Athens and elsewhere who sensed an injustice as the EU acted in its own interests at the expense of the crisis-hit southern European states. A confidential International Monetary Fund report leaked to the *Wall Street Journal* in June 2013 said that the EU had acted against the interests of Greece in 2010 in order to protect itself by building a financial firewall around the stricken economy. The report states that an immediate debt-restructuring programme would have benefited Greece more than the protracted response that exacerbated its financial downturn. However, such a response was rejected by

Greece's partners in the Euro because they did not want to destabilise either the single currency or their own national positions.[15] Furthermore, the EU's actions were described as 'a holding operation' that 'gave the euro area time to build a firewall to protect other vulnerable members.'[16] Early debt-restructuring, Greece's preferred option and a restructuring that eventually occurred in 2012 after Greece's austerity package had made its predicament worse and caused considerable hardship to individuals across the economy, 'had been ruled out by the euro area.'[17] The decision was defended robustly by the EU's Economics Commissioner Olli Rehn who said that an early debt-restructuring programme could have prompted an accidental breakup of the Euro [currency].[18] In his statement, Rehn thereby confirmed a key contention of this chapter and a central aspect of the truth wars that shape European politics: that maintaining what has been achieved so far and expanding further in the future is the EU's *raison d'être*. Institutional survival is its ultimate aim, a truth that cannot be challenged, or even be seen to be challenged, regardless of the cost to millions of individual citizens of EU member states like Greece.

In the preceding paragraphs a number of overlapping power relations and contested truth claims can be seen at work between different groups and interests: EU officials seeking at almost all costs to protect the institutions of the European Union, especially the Eurozone and its currency, the Euro; national leaders belatedly seeking to advance or protect the rights and privileges of the states they have been elected to represent; and individuals who form their identity in relation to both. The first dimension operates at state level: my identity as a Briton is partly dependent on my not being French, German, Austrian, and so on. Going further, my identity as a Briton can be formed in antipathy to those others – typically relying on a historical confrontation against which to define that otherness. The second dimension of otherness operates in relation to the European Union itself, its centralising of power eroding the sovereignty of the political community within which identity is constituted and self-constituted. As far as subjectivity is concerned, the ultimate goal of the European project is for individuals to form their identities primarily in the context of the EU as the pre-eminent political community: the subjugation or defeat of state-based nationalism marking the beginning of a 'true' European freedom. Freedom from the dangers of repeating a barbaric past with its violent power politics. A freedom which, perversely, consumes national freedoms – many of which have been fought for and guarded over centuries – in the process. The only problem is that increasing numbers of citizens are demanding freedom from the EU rather than freedom *in* the EU.

People in the member states most badly affected by the European element of the global financial crisis – Greece, Spain, Ireland, Cyprus – responded in a number of ways as they came to terms with the new economic realities which they widely perceived as being inflicted upon them. Resistance to the EU and the Euro currency hardened alongside anger towards national governments who were viewed by their people as complicit in the disaster that had befallen them. At an international level the financial crisis was met by structural and institutional realignments; at a subjective level the experience of millions of citizens was betrayal, disappointment, rage and political impotence. A common feeling of millions across the continent whose lives and livelihoods had been harmed as a result of decisions made by financial institutions and governments was that it was just *wrong*: a moral judgement based on the truth as they experienced it.

The EU as moral order: good vs. evil

The idea of a moral order that governs every aspect of an individual's life is not a new one, and it has been disputed by philosophers and theologians for millennia.[19] However, I reject the idea that there is some perfect, timeless and universal moral order that somehow shapes all societies, shared political life, individual ethical conduct and which merely needs to be uncovered and activated. Likewise, I also reject the idea of a perfect, timeless and universal truth somewhere that underpins such a moral order and which is just waiting to be discovered. Therefore, this part of the discussion will explore my claim that the European Union has come to see itself – and be seen by its supporters – as a kind of all-encompassing moral order to which its citizens should conform and in relation to which personal identity as a 'European' (as opposed to national identities) should emerge. I will then go on to look at how some in Europe (which is different to seeing oneself as 'European') are seeking to form or re-form their identities by rejecting the all-consuming, all-powerful EU and the seemingly unstoppable advance of its institutions into the lives of its citizens, opting for an individualised personal ethic instead.

The legacy of Augustine's fifth century writings remains strong in Europe, even in a post-Christian age. In his book *Confessions*, Augustine advanced the practice of reflecting on one's character, behaviour and beliefs, and confessing any and all shortcomings to the Christian God who would forgive the penitent soul. These ideas might appear at first glance to be completely disconnected from the experience of most people today but while there are definite differences there are also some highly significant parallels when it comes to shaping individual identity and conduct. Where Augustine practiced self-examination and confession of sin he did so in relation to an all-powerful, all-seeing God who held out the hope of eternity in heaven in the next life. In contrast, self-examination in the context of the recent and ongoing financial crisis is done so in relation to an all-too-earthly political power (the EU) that holds out the promise – or threat – of either heaven (ongoing membership) or hell (being cast out into isolated economic obscurity and all the dangers of an unprotected world), depending on one's perspective.

That the EU, its institutions, the Euro currency and the Eurozone where it operates are intrinsically good should be taken as the unquestioned fundament against which all decisions are and have been made and in relation to which Europeans are subjects. The moral order that defines and is defined by the idealised centralisation of European power and the erosion of individual state sovereignty, demands the conformity of every individual who is subject to it. To self-identify primarily as 'a European' is good; to self-identify as 'a Greek', 'an Italian', or 'a Briton' and so on, is bad. In the discourse of the European Commission or other institutions, to be – primarily – European is to be forward-looking, cosmopolitan, egalitarian, concerned for rights pertaining to gender, sexuality, race and religion. Most of all it is to commit oneself to a continual striving to escape the demons (and I use the word deliberately) of previous generations who wreaked unthinkable horror upon states and minority groups in a pursuit of power that began by redefining words like 'normal' and 'truth' and 'good', before culminating in the smoking chimneys of Auschwitz. In contrast, to identify primarily with the country of one's birth is presented as nationalist, inward and backward-looking, narrow minded and probably xenophobic.

The conduct of Switzerland illustrates the point. Although it is not a full member of the EU, numerous treaties between the two have allowed Switzerland to enjoy full economic access and trade benefits with the continental bloc in exchange for allowing the free flow of EU citizens to live and work there. In a referendum in 2014 the Swiss people voted by the narrow majority of 50.3 per cent to restrict immigration from the EU.[20] The following week, the EU's first practical response was to suspend talks on a major research-funding programme from which Switzerland was due to benefit.[21] On one level it is entirely appropriate that any political club, association or other collective should have the right to enforce the terms of its own existence. However, the absolutist approach of the EU supports my reading of it as not only a legal and political entity but as a self-protecting social order with its own moral framework and laws. The President of the European Commission, José Manuel Barroso, said categorically of one of the pillars of the moral (and legal) order: 'We will not negotiate on the principle of free movement.'[22]

Augustine identified and rooted out religious heretics who defied or rejected the rules and prohibitions set out in Catholic doctrine 1,600 years ago – rules that he helped to write. To be branded a heretic was to be ejected not only from the Church but also from the heavenly City of God. Similarly, aspects of the EU's existence and practices echo those found in the Christianity of its past. The principal European doctrine is that its existence cannot be called into question. If Augustine robustly enforced the Old Testament commandment that 'Thou shalt have no other gods before me', the EU and the Eurozone within it will not countenance any alternative to its continued existence. Anyone who suggests otherwise is by definition a heretic and branded a Eurosceptic, or worse. Consequently, when at the height of the financial crisis it was suggested that a member state like Spain or Greece could withdraw from the Eurozone, readopt its own currency, devalue that currency to make itself economically competitive once more, then slowly grow its way out of the crisis outside the Eurozone, the guardians of the Euro ensured that such talk was quickly silenced. Solutions to economic problems require individuals and member states to – in more ways than one – keep the faith in the European Union.

Such an account of political truth offers little comfort, or certainty, to the individuals who want to know why they and their families find themselves in financial dire straits. Surely someone, somewhere is at fault? Yet that uncertainty is mirrored in how the principal solutions to the economic crisis – financial austerity or financial stimulation – are offered. If there is such a thing as an underlying economic truth that will solve the economic problems that emerged in the trail of the financial collapse of 2008, then it should, in theory, simply be a case of identifying that truth and following where it leads. However, these competing truth claims (austerity or stimulation) are incommensurate and cannot be applied simultaneously in any one location (you could obviously apply them simultaneously in different economies). The dominant response to out-of-control debt in the Eurozone was to take measures to reduce that debt, such as austerity programmes and strict discipline when it comes to spending. Alongside the austerity programmes forced upon the southern European states as a condition for EU bailouts of their financial systems can be found contradictory advice from Christine Lagarde, head of the International Monetary Fund and herself a European, who has called for the easing of austerity.[23] It is in this confused and confusing context that individuals in badly affected countries try to regain a sense of who they are, individually and collectively, in light of what they have experienced – overwhelmed by uncertainties and with little hope that prosperity will return any time soon. Against this confusing backdrop, recourse to the language of good and evil, same and other, appeals to many: a return to the past that provides a sense – illusory or otherwise – of certainty in a world of chaos.

As the opening lines of this chapter implied, recalling a past world where Nazi Germany embodied all evils and giving it new life in the financial discourse of the present may bring temporary relief to those suffering as the institutions of the EU protect and preserve its existence. However, it could encourage isolationism which, taken to one ultimate conclusion, could result in an ethic of all against all. In September 2011 the Polish Finance Minister, Jasek Rostowski, made such a concern explicit when he warned that the European Union could not survive the breakup of the Eurozone and that war within Europe could be one possible outcome.[24] Rostowski's response is consistent with EU orthodoxy: there is one true path – accept the regime of European political truth – while the alternative is apocalyptic disaster – echoes of Augustine's choice between the joys of heaven or the trials of hell.[25] There is no suggestion from leading individuals in the EU or other European institutions that an alternative European polity might be desirable or successful; there is only the ideal of closer political and economic union or the terrors of isolationist oblivion. Put more simply: scare populations and governments into conformity if persuasion does not work.

In December 2012 the economist James Galbraith spoke of the risk of an 'explosion of violence' in Europe if struggling countries were allowed to proceed in an uninterrupted downward spiral. He quoted John Maynard Keynes, who warned in 1919 against allowing the humiliation of post-WWI Germany: 'The policy of reducing Germany to servitude for a generation, of degrading the lives of millions of human beings and of depriving a whole nation of happiness should be abhorrent and detestable. Abhorrent and detestable, even if it were possible, even if it enriched ourselves, even if it did not sow the decay of the whole civilized life of Europe.'[26] While the conditions which brought about Germany's ruin in 1918 and the conditions that brought about the financial crisis in Greece post-2008 are hugely different, Keynes's warning about the potential destructiveness of degrading millions of people seem strikingly relevant. Recall that the European Union, through its institutional predecessors, came into being to ensure that extreme nationalism of the sort that gave rise to Nazism during the inter-war years would not occur again. However, an unintended consequence of the EU's policies during the financial crisis has been resurgence in aggressive right-wing nationalism across several member states. The Greek general election in June 2012 saw a narrow majority vote for parties that supported the strict terms of the EU's bailout package. However, it also saw the far-right Golden Dawn Party receive 425,970 votes – almost 7 per cent of the national vote – and 18 seats in parliament.[27] One of those Golden Dawn MPs – Ilias Panagiotaros – highlighted its neo-Nazi tendencies by calling Hitler a 'great personality'.[28] Perversely, the EU has helped to incite that which it was created to prevent.

Ethics, identity and European politics

So how are people to form themselves as ethical European subjects in countries that have suffered several years of economic decline, rising unemployment, and when they see no end in sight to a financial torment that many believe was inflicted *upon* them rather than caused *by* them? I suggest there are two ways. The first is by conforming to the various codes – rules, values, laws, rights, institutional mechanisms and pronouncements, and federalist aspirations – that combine to create the European ideal as a form of moral order. The second is by pursuing a subjective ethic that is centred on individualised values, motivations, attitudes, goals and behaviours with national and cultural roots that are not necessarily based on the aforementioned codes.[29] Complicating matters – because they are invoked both by the codes that define European order and by individuals who oppose it – are conceptions of freedom and democracy. For example, a federalist might say that the only way to ensure freedom from tyranny for future generations is to pursue closer European integration. However, the opponent of federalism might argue that freedom and democracy only have meaning where everyone has the freedom to vote in elections to *every* European lawmaking body, supported by national referenda on whether or not they want that form of European integration. In every other part of the world where citizens are denied the right to vote into power those who would make the laws that govern their lives, such governments are described as authoritarian, tyrannical, anti-democratic or worse. In the EU it is easier for the concepts of 'freedom' and 'democracy' to simply be redefined and redeployed than to ask citizens to vote in elections that would make European institutions truly accountable. Consider an example of the complexities and contradictions thrown up in European political discourse when viewed from a 'bottom-up' perspective through the eyes of the individual rather than from a 'top-down' perspective of the European institutions.

The European Convention on Human Rights (ECHR) emerged from the Convention for the Protection of Human Rights and Fundamental Freedoms, first signed in Rome on 4 November 1950, which has its roots, in turn, in the Universal Declaration of Human Rights of 1948. The World War II Nazi Holocaust against (mainly) the Jews of Europe and the millions of civilian deaths caused by starvation, dislocation and modes of war such as area bombing engendered a global desire to avoid a repeat of such events in the future. Consequently the preamble to the ECHR states:

> The governments signatory hereto, being members of the Council of Europe ... Reaffirming their profound belief in those fundamental freedoms which are the foundation of justice and peace in the world and are best maintained on the one hand by an effective political democracy and on the other by a common understanding and observance of the human rights upon which they depend.[30]

In 1950 in Europe the scars of the previous decade ran deep; the desire to protect individual rights to life, liberty and security was an unquestioned and unquestionable good (and in my view, remains so). Subsequently, human rights were expanded into more and more areas of European public life and enacted in law throughout the EU. By 2012, however, such was the expansion of the ECHR into areas of public and private life unforeseen in 1950 that Prime Minister David Cameron was able to say to the Council of Europe about the European Court of Human Rights: 'at times it has felt to us in national governments that the "margin of appreciation" – which allows for different interpretations of the Convention – has shrunk and that not enough account is being taken of democratic decisions by national parliaments.'[31] His highlighting of the democratic deficit between the Convention and the people of Europe would appear to contradict the aims of the ECHR itself which refers to 'effective political democracy' as one of the conditions of protecting human rights.[32] Among many issues that led Cameron to call for changes to the European Court of Human Rights were frustrations about not being able to deport individuals that the UK government believed, or could prove, posed a threat to national security.

From the perspective of the individual, the democratic deficit looks even greater. Like most Britons I am a passionate believer in human rights and a supporter of the ECHR. However, I am also a strong supporter of democracy and in the 30 years since reaching voting age have never been offered a direct vote on any aspect of the UK's relationship with Europe. British and other governments have been encouraged to adopt a pro-European stance and actively dissuaded by leading figures in EU institutions from offering referenda on Europe. With regard to the European financial crisis the democratic deficit is even greater. Citizens of Greece, for example, had no real democratic choice about whether or not to accept the different bailout packages that their country needed. In one such rescheduling of Greek debts in 2012, painful terms were enforced upon the Greek government for a new EU/IMF bailout that had a catastrophic impact on the lives of millions of Greeks. Their only other option was abandoning the Euro and risking even greater national and personal disaster. The European Union and its remaining member states would have had no incentive to support a departing Greece: if it survived or even thrived outside the EU it would call into question the very need for that institution. It should come as little surprise that in such a position many Greeks increasingly saw the EU – and particularly the Germans whose funds were desperately needed – as the 'other' who would harm them rather than as beneficent fellow Europeans who wanted to help them.

Multiple tensions have been seen across Europe following the financial crisis as mutual respect between national political partners has been eroded or lost, with victimhood offering the most coherent expression of a new, emerging identity for many, often expressed in terms of anger, disappointment or betrayal. The historical inter-state rivalries that the European Union was created to ameliorate have morphed in some instances into state *versus* EU rivalries. Within those relations of power individuals are ethically implicated as they form new, hostile relationships with those who do not share their travails. What is the Greeks' ethical responsibility to the Germans and others who have bailed them out? In financial terms it has been to reform their indigenous practices: enforcing tax payments, raising retirement age, reducing retirement and other social benefits. However, the cost has been a huge shift in culture and historical practices. From the opposite perspective, to what extent is the German citizen morally responsible for bailing out the Spaniard, the Greek or the Cypriot and how does he or she emerge as ethical in doing so? While much debate on what measures to take is carried out at an intergovernmental level within the EU financial and other institutions, popular debate continues to emphasise the subjective – the actions that cause individual harm in some way. In terms of the German response, over the course of several years of economic crisis and multiple financial bailouts that have been funded predominantly by Germany, its people could be forgiven for becoming weary and increasingly resentful of the anger aimed at them by recipients of their help.

Respect, no matter how permissively the word is defined, is in chronically short supply between people and states that are referred to, without intended irony, as European partners. 'Partners' suggests some degree of equivalence in the relations of power that frame European politics but, when it comes to matters of finance and economy, the term is woefully inappropriate for the actual discourse that takes place. In terms of emerging personal identity, mutual respect has increasingly given way to mutual antipathy and a chronic breach of trust. If the launch of the Euro suggested that in the future, people in the member states would 'all be in it together' and that individual identities would be based on a homogenised relationship with the EU as the dominant political authority, the harsh, post-crisis realities of that shared outlook appear somewhat less than a wholesome mutuality. Consider a maritime analogy where all of the Eurozone states are sailing along reasonably happily on the 'Good Ship' Euro. At the moment the debt iceberg was struck it might be true to say that they were all in it together. However, with the Germans making the biggest contribution to the design, maintenance and running costs of the ship, at the moment of crisis it became immediately apparent that they own the 'Good Ship' Euro's lifeboats and lifebelts. Furthermore, they can decide who is to be rescued from the turbulent froth and the conditions that are to be placed on that rescue. Everyone may have been on the ship but some are much closer to drowning than others. Yet despite these events it is still unacceptable in EU political discourse to question the design and integrity of the ship in the first place.

For the Irish, Greeks, Spaniards and others whose economies have suffered as a consequence of the complex, imbalanced financial and juridical-political power relations on which the Eurozone was founded, difference has a new dimension that sits at the heart of personal identity and the way people view themselves. Twentieth-century history provides an easy source of anger and blame for the individuals who want to distance themselves from Germany in particular, complete with images of Nazi brutality and conquest, and historical myths of Aryan superiority. From Greece to Cyprus and beyond in the south of Europe, post-crash images of Angela Merkel superimposed with a Hitler moustache and wearing Nazi uniform evoke and parody a past that repulses, offends and fascinates at the same time – shorthand for ill-articulated and impotent rage,[33] photos that cry out: 'I might not now know who I am any more or how I got here but I know who I blame: YOU.'

The complaints are about financial collapse, poverty, youth unemployment, contracting economies, and much more besides. While all of this seems clear it still does not provide a satisfactory explanation. Loss is not only expressed in terms of finance and democracy, the loss extends to personal and shared identity. For those struggling to make sense of their place in this new, austere world the other is not only the apparently dominant, successful Germany and its people, the other is also the self that once was: the Greek, the Cypriot, the Spaniard of the carefree and seemingly prosperous days before an imploding financial market turned the world upside down for tens of millions of Europeans. When, in 2011, Volker Kauder told Germany's governing Christian Democratic party's annual conference, 'Suddenly Europe is speaking German',[34] what may have been intended as a practical statement of economic necessity was interpreted as much more sinister and threatening by many who would prefer to keep speaking their native languages. Personal politics became increasingly framed by anxiety, loss, anger, blame and victimhood, resulting in the ethical dimension of a shared European existence being gravely undermined.

Central to this truth and morality discourse, even in Western post-Christian societies where such language is otherwise derided as tired and out-of-date, is opposition to evil: whether that be the evil of historical Christianity, the more recent but all-too-human barbarism of the concentration camp, or the contemporary destruction of hope in the everyday lives of millions. The ideals on which the European Union was founded provide little comfort when human dignity is violated, when proud men and women beg for their country to be rescued, their anger so focused on blaming others that personal and collective responsibilities are overlooked.

And what of the rescuer? Decades of personal and national fiscal propriety – even absorbing the monumental cost of reunification of East and West Germany – allied to economic expansion, has brought Germany to its current position of political and economic dominance in Europe. Deliberate avoidance of remilitarisation has been the backbone of Germany's defence policy since World War II and yet the voice of the Chancellor – no matter how quietly she whispers – has become the loudest on the continent. In May 2013, Angela Merkel was discussing Hungary's future in the European Union, its contracting economy, and concerns about recent constitutional changes that threatened to damage democratic legitimacy. She commented that Berlin would 'do anything to get Hungary onto the right path – but not by sending the cavalry'.[35] In response, Hungary's prime minister, Viktor Orbán, replied in a radio interview: 'The Germans have already sent cavalry to Hungary – they came in the form of tanks...Our request is that they don't send any. It didn't work out.'[36]

Summary

The European Union came into existence as a result of a desire to unify the people of Europe with the aim of avoiding a repeat of the catastrophes of 1914–1918 and 1939–1945. A crucial element of the European project was to reduce differences and alienation between peoples, the same 'difference' that plays a key role in the forming of personal identity. However, the focus on 'top-down' institutional responses to the economic crisis that befell Europe, and many other parts of the world, from 2008 onwards frequently ignored or downplayed the consequences for individuals and their personal and collective identities. The priority was to get the frameworks right, the laws in place, and the institutions more efficient to provide stable governance. The emergence of a common currency to unite historical enemies in the common cause of economic prosperity and shared success has had the unintended consequence of restoring and magnifying some of the very differences that it had been introduced to eradicate. Furthermore, the formation of personal identity in relation to a common Europe rather than in relation to the narrower confines or preferences of traditional nationality has similarly reached its limits during the post-2008 financial crisis.

Several years later, with ongoing economic challenges still being faced, nationalism is growing across many, perhaps most, European states: the antithesis of the original goals of the champions of a united Europe. Difference is being increasingly stressed on a continental scale as shared identity recedes back into historically familiar geopolitical domains. Consequently, to be Greek is *not* to be German and increasingly *not* to be European in the sense of a united people. In the truth wars, it has been increasingly perceived that 'Europe' is what richer northern states are doing to, or inflicting upon, economically fragile southern states, with the former holding the dominant position in the relations of power that define policy decisions to provide financial stability to rich and poor alike. That Chancellor Angela Merkel and President François Hollande should hold a Paris summit to address 'the revival of Europe'[37] indicates the extent to which its original ideals have been buffeted by the economic crisis. As the next stage of the European project is being planned, a warning from Keynes in 1919 retains a striking relevance almost a century later: 'The bankruptcy and decay of Europe, if we allow it to proceed, will affect everyone in the long-run, but perhaps not in a way that is striking or immediate.'[38] It might not be polite to mention the war but it may also provide a helpful corrective to remind the world how bad things can get if antipathy and anger are set free from the constraints of trust, respect and mutual concern. It might also be helpful to consider that there might be more nuanced options available than a simplistic choice between good and evil, between ever-closer European political union and scare stories of apocalyptic nationalism.

10
Epilogue

A book such as this cannot rightly end with a Conclusion. To do so would imply a degree of finality that is not appropriate when the truth wars surrounding climate change, unfolding tragedies that might prompt new military interventions, and aftershocks from the financial crisis are ongoing and continue to shape individual lives and global politics. Circumstances change, new crises emerge and priorities shift rapidly in a turbulent world.

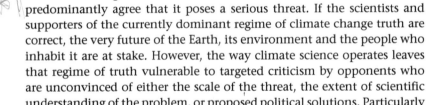

Climate change remains as contentious as ever and climate scientists predominantly agree that it poses a serious threat. If the scientists and supporters of the currently dominant regime of climate change truth are correct, the very future of the Earth, its environment and the people who inhabit it are at stake. However, the way climate science operates leaves that regime of truth vulnerable to targeted criticism by opponents who are unconvinced of either the scale of the threat, the extent of scientific understanding of the problem, or proposed political solutions. Particularly where any proposed solutions can seem motivated more by a desire to redistribute global wealth, have developed nations atone for historical 'wrongs' against their developing counterparts, discourage future economic growth and the advance of capitalism, or promote a universal, simple egalitarian lifestyle, rather than ameliorate specific climate-related problems. That ideological battle will rage most fiercely of all.

The scientific content of the IPCC 5th Assessment Report 'Climate Change 2014: Mitigation of Climate Change' was accepted by the IPCC on 12 April 2014, prompting early political machinations in advance of the 2015 meeting of members of the UN Framework Convention on Climate Change. Government delegates then selected elements of the scientific report and decided on the emphasis that specific parts would or would not receive in the Summary for Policymakers.[1] They may have started with a scientific document but they finished with a political compromise. Truth wars in action as developed, developing and poor countries alike began to lay the foundations for the claims they would be making against everyone else in international negotiations in 2015.

In addition, when so many scientists see climate change as a 'cause' there is a clear potential to undermine – and occasionally call into disrepute – the scientific principles of universalism, communalism, disinterestedness and organised scepticism. And all of this taking place against the backdrop of a 16-year 'pause' in the rise of global mean temperatures. A pause measured by NASA, the US National Climatic Data Center (NCDC), and the UK Met Office; a pause that is lionised by opponents of the regime of climate truth; and a pause that is denied by key figures at the IPCC. Perversely, climate change opponents are basing their arguments on the temperature measurements of climate scientists (the 'pause') while other pro-climate change scientists deny or downplay the significance of the measurements. Meanwhile, climate Armageddon seems no nearer than it was 30 years ago and interested, open-minded lay observers are still waiting to hear why the measured early twenty-first century 'pause' in rising global mean temperature was not predicted loudly and clearly in the 1980s and 1990s, and what its implications are now, if any.

What is currently beyond dispute is the downgrading of the climate threat by governments in most major developed and developing economies. The global financial crisis has exposed the extent to which tackling climate change has fallen down the list of political priorities. Jobs, economic growth, standards of living and energy security have explicitly been prioritised from Washington to London to Canberra and every continent in between. At the risk of future embarrassment I am willing to predict that it will take a rapid, severe spike in global warming coupled with an extraordinary (by the standards of the last several hundred thousand years, not just the last two centuries since industrialisation) series of climate-related events to give climate change the political clout and public support it enjoyed at the turn of the twenty-first century.

In parallel to the recent climate truth wars, military intervention as a means of advancing Western liberal democracy is reaching its denouement. Recent events in Iraq, Libya and Afghanistan have demonstrated the difficulty of using military force as a means of extending liberal democracy and associated conceptions of freedom to people and cultures who have not sought it for themselves. Further, they have shown the limits of the financial and human sacrifice that the American and British people, together with their NATO allies, will endure to try to extend their social, political and ethical values to others who prioritise alternative cultural and religious truths. History will show that the liberal interventionist period – where interventions were motivated by the advancement of specific values and notions of freedom and democracy – ran from 1999 to 2011 and foundered on the rock of the subsequent Syria crisis. Future interventions may be prompted by human catastrophe but Western responses will be increasingly shaped by *realpolitik* and national self-interest: restrained by weary and wary populations. Nascent democracy – if it can even be called that – in Iraq, Libya and Afghanistan is a long way from being sustainable, at least in the form envisaged by the authors of those interventions. So many of the optimistic words about opposing tyranny and promoting freedom and democracy by Tony Blair, Bill Clinton, George W. Bush, Barack Obama and David Cameron now seem misguided at best and deluded at worst.

'Freedom' in the countries that have received the dubious benefit of US and UK intervention will be redefined and enacted by the populations concerned. Initially it will be freedom *from* the US and UK that matters most, followed by the implementation of culturally and religiously appropriate definitions of the word. As Afghan women are increasingly repressed once more – by Western standards – other truth claims will increasingly come to the fore. Taliban and other religious leaders already refer to Muslim women being 'freed' from Western-style sexualisation and exploitation, 'freed' instead to conform to the demands of the Koran (and the demands of their fathers, husbands and brothers). If those women are to enjoy anything like Western freedoms in the long term a huge price will have to be paid. It is not a foregone conclusion that the Taliban and some of the more extreme tribal groups will once more dominate public life in Afghanistan after US, UK and other NATO troops withdraw in 2014. However, great sacrifices will have to be made if Afghans are to avoid a return to the dark ages of the 1990s. These are dark truth claims that I hope will be proved to be unfounded.

Elsewhere, in the truth claim and counter-claim concerning military intervention and America's ongoing war against Al-Qaeda – and it is still a war – the drone wars are anomalous. Where governments have been quick to engage with the public over the justification of the use of force in pursuit of democracy, freedom and human – especially women's – rights, the American and British governments' approaches to the use of the Reaper and Predator drones have been characterised by their silence. While drone strikes in themselves are a form of discourse, public engagement by those who command their use – political leaders – has been avoided or minimised for as long as possible. Consequently, and unusually, public debate and the terms of public debate (for example, using the meaning-laden 'drone' instead of UAV or RPA[2]) have come to be set by anti-drone campaigners and organisations. If American and British governments want to have the option to use drones in the future to pursue their policy and security ends, greater engagement with their respective publics will be needed. The potential benefits of persistence of surveillance and intelligence-gathering capability, together with the accuracy and proportionality of lethal strikes when required, will be lost if the technology is abused or if the general public loses confidence that they are not being used for secret and nefarious purposes.

Finally, turning once more to the global economy, unanswered questions remain. For example, why, several years after the financial crisis, are so many safeguards still to be put in place? The answer goes right to the nexus of democratic politics, the financial truth wars that continue to rage, and the extent to which day-to-day political reality reflects the idealism of one-person-one-vote. On the one hand, in the US, UK and elsewhere there is overwhelming public support for constraining the activities of banks and bankers, prompted by deep-seated mistrust of the individuals and institutions concerned.[3] Yet lawmakers struggle to put into place the rules and codes that will prevent – or at least ameliorate – a future disaster. If democratic mandate could be automatically translated into policy making there would have been a greater number of more stringent regulations put in place already. On the other hand, banks continue to wield huge financial power, despite so many of them having had to rely on their governments to either bail them out directly or pump sufficient liquidity into the financial system that it provided breathing space for the strugglers to pull through. In the financial truth wars the banks and the powerful few that control them continue to wield sufficient power to have thus far prevented the kind of constraints being imposed that popular opinion demands.[4]

Several leading economies are returning to at least slow economic growth, with the British economy forecast to perform strongly in 2014. However, the towering debts that funded the rescue of the global banking system in 2007–2008 continue to rise, not fall. A potential crisis point looms when the Fed and other central banks stop pumping cash into their national banking systems. A second potential flashpoint will occur when interest rates are raised from their current historic low levels. Many individual and corporate debts may prove to be unsustainable. Another round of mortgage and business defaults could prove problematic because the big economic lever of liquidity injection has already been used. Plus, the debt still remains from the last time: nobody knows what effect a second instalment of national debt acquisition would have but it would not be positive for the average citizen.

Finally, unforeseen political manoeuvring can unexpectedly open up new, combined fronts in the truth wars. Consider the destabilisation of Ukraine and Crimea's return to Russian rule: annexation by an aggressive power or a voluntary return to its historic home based on democratic mandate, depending on one's perspective. The US, EU and NATO continue to debate long and hard about what actions to take to prevent other parts of Ukraine, or the whole country, being taken over by Russia. In parallel, President Putin is determined to stop what he sees as the aggressive advance of NATO towards Russia's border. He also declares that he wants to protect ethnic Russians and Russian speakers within Ukraine. Actions proposed by the US, UK, Germany, France and Italy fall far short of military intervention as a means to ensure ongoing freedom for Ukraine and its citizens. In addition, Russia's threat to use its energy supplies to Europe as an economic coercive weapon has prompted the EU to explore ways of reducing its dependence on Russian gas. The partial return to using traditional, dirty coal is being proposed as a means of reducing that dependence on Russian energy. However, the associated secondary effect of increasing CO_2 levels in the atmosphere will necessarily undermine the regime of climate truth and its ecological priorities. Furthermore, any serious level of economic dispute between East and West can only harm the economies of both, at a time of financial fragility. The G8 has, at least temporarily, become the G7 after Russian exclusion. At every stage, truth claims are being made and contested, with the search for stability perhaps being the most seductive and unsustainable truth of all in the never-ending series of crises that characterise global politics.

Notes

Introduction

1. From the title of his book Harold D. Lasswell, *Politics: Who Gets What, When, How* (Michigan: Whittlesey House, 1936).
2. Michel Foucault, *Society Must Be Defended: Lectures at the Collége de France, 1975–76* (New York: Picador, 2003) p. 15.
3. Paris Welch, interview, available at: http://www.msnbc.msn.com/id/28001417/ns/business-stocks_and_Economy/t/bush-administration-ig-nored-clear-warnings/, accessed 20 June 2012.
4. Michel Foucault, 'Truth and Power', in Colin Gordon (ed.), *Power/Knowledge* (New York: Pantheon Books, 1980) p. 131.
5. For a serious foray into the theory of truth, and an introduction to some of the debates therein, I recommend Simon Blackburn and Keith Simmons (eds) *Truth* (Oxford: Oxford University Press, 2010).
6. Alfred Tarski, 'The Semantic Conception of Truth and the Foundations of Semantics', *Philosophy and Phenomenological Research*, 4 (1944), in Simon Blackburn and Keith Simmons (eds), *Truth* (Oxford: Oxford University Press, 2010) p. 118.
7. Tarski, in Blackburn and Simmons, 2010, p. 118.
8. Tarski, in Blackburn and Simmons, 2010, p. 118.
9. The discerning reader will already have noted my debt to the French philosopher and cultural historian Michel Foucault, especially with regard to his ideas about the inter-relatedness of truth, power, and the ways that individual identity – he prefers the broader term subjectivity – is shaped, especially with regard to ethics. See Michel Foucault, *The History of Sexuality Volume 2: The Use of Pleasure*, Trans. R. Hurley, (London: Penguin Books, 1984); or Michel Foucault, *The Hermeneutics of the Subject: Lectures at the Collège de France 1981–1982*, F. Gros (ed.) Trans. G. Burchell, (New York and Basingstoke: Palgrave Macmillan, 2001).
10. Although known as the Patriot Act, its full title is *Uniting and Strengthening America by Providing Appropriate Tools Required to Intercept and Obstruct Terrorism (USA Patriot Act) Act of 2001*, Public Law 107–56–26. 26 October 2001, available at: http://www.gpo.gov/fdsys/pkg/PLAW-107publ56/pdf/PLAW-107publ56.pdf, accessed 12 December 2013.
11. Alexander King and Bertrand Schneider, *The First Global Revolution: A Report by the Council of the Club of Rome* (Orient Longman, 1991) p. 75. http://www.geoengineeringwatch.org/documents/TheFirstGlobalRevolution_text.pdf

1 Climate, Science and Truth

1. The hiatus in temperature increase has been identified by the three internationally recognised centres that compile Global-average temperature

records: Goddard Institute for Space Studies (GISS), which is part of NASA (USA); National Climatic Data Center (NCDC), which is part of the National Oceanic and Atmospheric Administration (NOAA) (USA); and the UK Meteorological Office, in collaboration with the Climatic Research Unit (CRU) at the University of East Anglia (UK). See http://www.metoffice.gov. uk/climate-guide/science/temp-records, accessed 9 January 2014. Further, the nuances of this claim are explored at length by the London School of Economics' Grantham Research Institute on Climate Change and the Environment, and demonstrates how statistics can be used in multiple competing ways – 'Anthropogenic Global wW "Stopped" in 1997 ... and in 1996, 1995, 1982, 1981, 1980, 1979, 1978 and 1972', http://www.lse. ac.uk/GranthamInstitute/news/anthropogenic-global-warming-stopped-in-1997and-in-1996–1995–1982–1981–1980–1979–1978-and-1972/

2. Karl R. Popper, *Objective Knowledge: An Evolutionary Approach* (Oxford and New York: Oxford University Press, 1979) p. 342.
3. Paul Feyerabend, *Against Method*, 3rd Edition (London and New York: Verso, 2008) p. 238.
4. National Academy of Sciences, *On Being a Scientist: A Guide to Responsible Conduct in Research*, 3rd Edition (Washington: The National Academies Press, 2009) p. x.
5. Robert K. Merton, 1942, 'The Normative Structure of Science', in Norman W. Storer (ed.) *The Sociology of Science: Theoretical and Empirical Investigations* (Chicago: University of Chicago Press, 1973) pp. 267–278.
6. Thomas S. Kuhn, *The Structure of Scientific Revolutions*, 3rd Edition (Chicago and London: The University of Chicago Press, 1996) p. 35.
7. BBC News, 4 July 2012, 'Higgs Boson-like Particle Discovery Claimed at LHC', http://www.bbc.co.uk/news/world-18702455, accessed 30 December 2012.
8. The Atlas Collaboration, 'A Particle Consistent with the Higgs Boson Observed with the ATLAS Detector at the Large Hadron Collider', *Science*, Vol. 338 (2012) pp. 1581–1582.
9. Peter Higgs, 6 July 2012, interview at Edinburgh University, footage located at http://www.guardian.co.uk/science/video/2012/jul/06/peter-higgs-edinburgh-university-video?INTCMP=ILCNETTXT3486, accessed 24 July 2012.
10. Andrew Dessler and Edward A. Parson, *The Science and Politics of Global Climate Change*, 2nd Edition (Cambridge: Cambridge University Press, 2010) p. 36–37.
11. Dessler and Parson, *Global Climate Change*, p. 37.
12. Stillman Drake, *Galileo At Work* (Chicago: University of Chicago Press, 1978) p. 367.
13. The concept of a regime of truth is used here in the sense set out in the Introduction.
14. Barry J. Marshall and J. Robin Warren, 'Unidentified Curved Bacilli in the Stomach of Patients with Gastritis and Peptic Ulceration', *The Lancet*, Vol. 323, Issue 8390 (June 1984), pp. 1311–1315.
15. Thomas S. Kuhn, *The Structure of Scientific Revolutions*, 3rd Edition (Chicago and London: The University of Chicago Press, 1996) p. 166.
16. Kuhn, *Scientific Revolutions*, p. 10.
17. Kuhn, *Scientific Revolutions*, p. 175.
18. Stephen H. Schneider, October 1989 interview with *Discover* magazine, reprinted in Detroit News Editorial Response, 5 December 1989, http://

stephenschneider.stanford.edu/Publications/PDF_Papers/DetroitNews.pdf, accessed 17 December 2012.

19. Michael E. Mann, *The Hockey Stick and the Climate Wars* (New York and Chichester: Columbia University Press, 2012) p. 26.
20. Mann, *Hockey Stick*, p. 76–77.
21. James Delingpole, *Watermelons: How Environmentalists are Killing the Planet, Destroying the Economy and Stealing Your Children's Future* (London: Biteback Publishing, 2012) p. 104.
22. Robert M. Carter, *Climate: The Counter Consensus* (London: Stacey International, 2010) p. 150.
23. Merton, 'Normative Structure of Science'.
24. Schneider, October 1989 interview.
25. Schneider, October 1989 interview.
26. Schneider, October 1989 interview.
27. Silvio O. Funtowicz and Jerome R. Ravetz, 'Science for the Post-Normal Age', *Futures*, Vol. 25, No. 7 (1993) 739–755. See also discussion in Mike Hulme, *Why We Disagree About Climate Change* (Cambridge: Cambridge University Press, 2011) p. 78.
28. Silvio O. Funtowicz and Jerome R. Ravetz, 2003, 'Post-Normal Science', *International Society for Ecological Economics*, p. 3, http://leopold.asu.edu/ sustainability/sites/default/files/Norton, per cent20Post per cent20Normal per cent20Science, per cent20Funtowicz_1.pdf accessed 10 December 2013.
29. Funtowicz and Ravetz, 'Post-Normal Science', p. 4.
30. Gert Geominne, 'Has Science Ever Been Normal? On the Need and Impossibility of a Sustainability Science', *Futures*, No. 43 (2011) p. 628.
31. Geominne, 'Has Science Ever Been Normal?' p. 631.
32. Geominne, 'Has Science Ever Been Normal?' p. 632.
33. Dessler and Parson, *Global Climate Change*, p. 41.
34. Dessler and Parson, *Global Climate Change*, p. 39 (my emphasis).
35. Randy Schekman, 9 December 2013, 'How Journals Like Nature, Cell and Science are Damaging Science', http://www.theguardian.com/ commentisfree/2013/dec/09/how-journals-nature-science-cell-damage-science, accessed 19 December 2013.
36. Schekman, 'How Journals Like Nature, Cell and Science are Damaging Science'.
37. For contrasting perspectives on the detail and implications of 'Climategate' (a term popularised by Delingpole), see James Delingpole, *Watermelons*, pp. 17–40; Michael E. Mann, *The Hockey Stick, and the Climate Wars* (New York and Chichester: Columbia University Press, 2012) pp. 207–232.
38. Willie Soon and Sallie Baliunas, 'Proxy Climatic and Environmental Changes of the Past 1000 Years', *Climate Research*, Vol. 23 (2003) pp. 89–110.
39. Email no. 104738848 from Michael Mann to Phil Jones, 11 March 2003, located at Mann to Phil Jones, 11 March 2003, located at http://www.ecowho. com/foia.php?file=1047388489.txt&search=The+Soon+%26+Baliunas+paper accessed 19 July 2014.
40. George W. Bush, 20 September 2001, Transcript of President Bush's Address to a Joint Session of Congress and the Nation, located at http://www.washington-post.com/wp-srv/nation/specials/attacked/transcripts/bushaddress_092001. html, accessed 6 April 2014.
41. Mann, *The Hockey Stick*, p. 96–97.

42. Mann, *The Hockey Stick*, p. 79.
43. Email from Steve Mackwell to Michael Mann, 20 January 2005, http://www.ecowho.com/foia.php?file=2065.txt&search=Steve+Mackwell accessed 19 July 2014.
44. Circular email from Michael Mann, 20 January 2005.
45. Delingpole, *Watermelons*, p. 16.
46. Carter, *Climate: The Counter Consensus*, p. 54.
47. Schneider, October 1989 interview.

2 Politics and Climate Truth

1. Ed Davey, 13 February 2014, Speech to the Institute for Public Policy Research: 'Energy Divided? Building Stability in Energy Policy', http://www.libdemvoice.org/ed-daveys-speech-to-the-ippr-on-climate-change-38220.html, accessed 7 April 14.
2. Hulme, *Why We Disagree about Climate Change*, pp. 361–362.
3. Hulme, *Why We Disagree about Climate Change*, pp. 362.
4. Hulme, *Why We Disagree about Climate Change*, pp. xxix–xxxiii.
5. Carter, *Climate: The Counter Consensus*, p. 47 (my emphasis).
6. John Beddington, 25 March 2013, television interview on BBC Breakfast News.
7. Beddington, 25 March 2013.
8. Michel Jarraud, http://www.theguardian.com/environment/2014/mar/31/climate-change-report-ipcc-governments-unprepared-live-coverage, accessed 7 April 2014.
9. Mann 2012, p. 254.
10. Mann 2012, p. 256.
11. The Climatic Research Unit is part of the University of East Anglia and the UK's most high profile and influential climate research centre.
12. Michael Mann, 3 August 2004, email to Phil Jones located at http://www.ecowho.com/foia.php?file=3115.txt&search=03%2F08%2F2004, accessed 19 July 2014, (my emphasis).
13. Peter Thorne, http://globalchange.ncsu.edu/march-25-a-conversation-with-peter-thorne/, accessed 19 July 2014.
14. Peter Thorne (2005), private email to Phil Jones, 20 February 2005, located at http://www.ecowho.com/foia.php?file=3066.txt&search=20%2F02%2F2005, accessed 19 July 2014. Author: This link now works.
15. Thorne, private email to Phil Jones, 20 February 2005.
16. Thorne, private email to Phil Jones, 20 February 2005.
17. Dessler and Parson, 2010, p. 39.
18. Phil Jones, 21 February 2005, private email to Warwick Hughes, located at http://www.ecowho.com/foia.php?file=1299.txt&search=Mon+Feb+21accessed 19 July 2014.
19. Phil Jones, 2 February 2005, private email to Michael Mann, located at http://www.ecowho.com/foia.php?file=1107454306.txt&search=2%2F2%2F2005 accessed 19 July 2014.
20. Delingpole, *Watermelons*, p. 36.
21. Phil Jones, 11 March 2004, email located at http://www.ecowho.com/foia.php?file=4443.txt&search=Thu+Mar+11 accessed 19 July 2014.

22. Christopher Monckton, 27 December 2006, email located at http://www.ecowho.com/foia.php?file=0112.txt&search=05+Feb+2007, accessed 19 July 2014.
23. Heliogenic: produced by, or relating to, the Sun.
24. Michael Mann, 5 February 2007, email located at http://www.ecowho.com/foia.php?file=0112.txt&search=05+Feb+2007, accessed 19 July 2014. Author: This link now works.
25. Mann, *The Hockey Stick*, p. 257.
26. Mann, *The Hockey Stick*, p. 257.
27. Mann, *The Hockey Stick*, p. 257.
28. Michel Foucault, *Power/Knowledge: Selected Interviews and Other Writings by Michel Foucault, 1972–1977* (New Pork: Pantheon Books, 1980) p. 131 (my emphasis).
29. Peter Sissons, *Daily Mail*, 9 February 2011.
30. Peter Sissons, *Daily Mail*, 9 February 2011..
31. BBC Charter, 28 June 2006, 'Regulatory Obligations on the UK Public Services', *Broadcasting: An Agreement Between Her Majesty's Secretary of State for Culture, Media and Sport and the British Broadcasting Corporation* (London, The Stationery Office, 2006) p. 20ff.
32. BBC Trust Report, June 2007, 'From Seesaw to Wagonwheel: Safeguarding Impartiality in the 21st Century' p. 40, located at http://www.bbc.co.uk/bbctrust/assets/files/pdf/review_report_research/impartiality_21century/report.pdf, accessed 20 July 2012.
33. Richard D. North, 14 December 2008, email to Tony Newbery, located at http://ccgi.newbery1.plus.com/blog/?p=142, accessed 27 December 2013.
34. A full list of names can be found at http://omnologos.com/full-list-of-participants-to-the-bbc-cmep-seminar-on-26-january-2006/, accessed 27 December 2013.
35. BBC Trust Report, June 2007, p. 40.
36. Mann, *The Hockey Stick*, p. 257.
37. Gill Ereaut and Nat Segnit, *Warm Words: How Are We Telling the Climate Story and Can We Tell It Better?* (London: Institute for Public Policy Research, 2006) p. 8.
38. Gill Ereaut and Nat Segnit, *Warm Words*, p. 5.
39. http://www.ippr.org/about-us, accessed 20 December 2013.
40. http://www.ippr.org/about-us, accessed 20 December 2013.
41. http://www.ippr.org/about-us/how-we-are-funded, accessed 21 December 2013.
42. Schneider, October 1989 Interview.
43. This is not an attempt to achieve dramatic effect, I am actually stopping.
44. The paleoclimatologist Robert Carter will be viewed by many global warming advocates as a contrarian or denier because he argues that there is nothing exceptional about recent climate variation when examined in light of the millions of years of Earth's climate history. It is ironic that he acknowledges that 'abrupt climatic changes' have happened thousands of years ago – over as little as three years – and are likely to reoccur. See Carter, *Climate: The Counter Consensus*, pp. 26, 43.
45. *IPCC Fourth Assessment Report: Climate Change 2007*, 10.5.4.6 Synthesis of Projected Global Temperature at Year 2100, located at http://www.ipcc.

ch/publications_and_data/ar4/wg1/en/ch10s10–5-4–6.html, accessed 10 December 2013.

46. K. Humphrey, 2009, private email to Phil Jones, 19 May 2009, located at http://www.ecowho.com/foia.php?file=2495.txt&search=19+May+2009, accessed 19 July 2014. Author: This link now works.

47. *IPCC Fourth Assessment Report: Climate Change 2007*, The Himalayan Glaciers, 10.6.2. http://www.ipcc.ch/publications_and_data/ar4/wg2/en/ch10s10–6-2. html, accessed 29 December 2013.

48. Vijay Raina, 2009, 'Himalayan Glaciers – A State-of-Art Review of Glacial Studies, Glacial Retreat and Climate Change', Ministry of Environment & Forests, Government of India, http://moef.nic.in/sites/default/files/MoEF per cent20Discussion per cent20Paper per cent20_him.pdf accessed 25 April 2013.

49. Vijay Raina, 2009, 'Himalayan Glaciers', p. 7.

50. Rajendra Pachauri, Interview on New Delhi Television Ltd, 9 November 2009, footage located at http://www.youtube.com/watch?v=bnYmQjFoNCs, accessed 20 November 2012.

51. Pachauri, Interview on New Delhi Television.

52. Pachauri, Interview on New Delhi Television.

53. IPCC, 20 January 2010, 'Statement on the Melting of Himalayan Glaciers', located at http://www.ipcc.ch/pdf/presentations/himalaya-statement-20january2010.pdf, accessed 1 September 2012.

3 One World, Two Visions

1. United Nations Framework Convention on Climate Change, Kyoto Protocol, http://unfccc.int/kyoto_protocol/items/2830.php, accessed 15 December 2013.

2. http://www.bbc.co.uk/news/science-environment-25383373, accessed 17 December 2013.

3. For examples see: *The Australian*, http://www.theaustralian.com.au/opinion/ editorials/stuck-on-a-ship-of-cold-fools/story-e6frg71x-1226793309195, accessed 2 January 2014; and *The Wall Street Journal*, http://online.wsj.com/ news/articles/SB10001424052702304591604579292611684898656, accessed 2 January 2014.

4. http://www.theguardian.com/environment/2014/mar/31/climate-change-report-ipcc-governments-unprepared-live-coverage, accessed 6 April 2014.

5. Lorraine Whitmarsh, 'What's in a Name? Commonalities and Differences in Public Understanding of "Climate Change" and "Global Warming",' *Public Understanding of Science*, Vol. 18 (2009) 401.

6. Malay chicken fried rice.

7. Pope Francis I, 19 March 2013, 'Homily for Inaugural Mass of Petrine Ministry,' Http://Www.News.Va/En/News/Pope-Homily-for-Inaugural-Mass-of-Petrine-Ministry, Accessed 20 March 2013.

8. Pope Francis I, 19 March 2013.

9. Anthropocentric: human centred.

10. Lynn White, 'The Historical Roots of our Ecologic Crisis [with discussion of St Francis; reprint, 1967],' *Ecology and Religion in History*, (New York: Harper

and Row, 1974), p. 4, http://www.uvm.edu/~gflomenh/ENV-NGO-PA395/articles/Lynn-White.pdf, accessed 1 January 3014.

11. White, 'The Historical Roots of Our Ecologic Crisis,' p. 4.
12. For a discussion of consumer-driven Christianity in the US see John Micklethwait and Adrian Woolridge, *God is Back: How the Global Rise of Faith is Changing the World* (London and New York: Penguin Books, 2009).
13. *2011 Yearbook of American and Canadian Churches*, cited at http://www.ncccusa.org/news/110210yearbook2011.html, accessed 1 January 2014.
14. White, 'The Historical Roots of Our Ecologic Crisis,' p. 2.
15. Pope John Paul II, 1 January 1990, Message for the Celebration of the World Day of Peace, http://www.vatican.va/holy_father/john_paul_ii/messages/peace/documents/hf_jp-iI_mes_19891208_xxiii-world-day-for-peace_En.html, accessed 1 January 2014.
16. Pope John Paul II, 1 January 1990.
17. Pope John Paul II, 1 January 1990.
18. Joel Osteen, *Your Best Life Now* (New York: Faith Words, 2004) p. 5.
19. Stephanie Kaza, *Mindfully Green: A Personal and Spiritual Guide to Whole Earth Thinking* (Boston & London: Shambhala Publications: 2008) p. 7.
20. Kaza, *Mindfully Green*, p. 35.
21. Edward R. D. Goldsmith, Editorial, *The Ecologist*, Vol. 1, No. 1 (July 1990) p. 3.
22. Goldsmith, *The Ecologist*, p. 4.
23. Goldsmith, *The Ecologist*, p. 4.
24. Goldsmith, *The Ecologist*, p. 5.
25. Goldsmith, *The Ecologist*, p. 5.
26. Goldsmith, *The Ecologist*, p. 5.
27. Alexander King and Bertrand Schneider, *The First Global Revolution: A Report by the Council of the Club of Rome* (Orient Longman, 1991) p. 75.
28. Tony Blair, 26 March 2006, Statement to the Climate Change and Governance Conference, Wellington, located at http://tna.europar-chive.org/20061101025041/http://www.fco.gov.uk/servlet/Front/TextOnly?pagename=OpenMarket/Xcelerate/ShowPage&c=Page&cid=11072 98302322&to=true accessed 3 January 2014.
29. UN Security Council Debate on the Impact of Climate Change, 17 April 2007, located at http://www.un.org/News/Press/docs/2007/sc9000.doc.htm, accessed 19 March 2013.
30. UN Security Council Debate on the Impact of Climate Change, 17 April 2007.
31. UN Security Council Debate on the Impact of Climate Change, 17 April 2007.
32. UN Security Council Debate on the Impact of Climate Change, 17 April 2007.
33. UN Security Council Debate on the Impact of Climate Change, 17 April 2007.
34. There is extensive literature on Peak Oil but Richard Heinberg summed up the arguments effectively in *The Party's Over: Oil, War and the Fate of Industrial Societies* (Gabriola Island, BC: New Society, 2003).
35. Richard Heinberg, *The End of Growth: Adapting to Our New Economic Reality* (Gabriola, Clareview, 2011) pp. 17–18.
36. The Climate Change Act, 2008, http://www.legislation.gov.uk/ukpga/2008/27/contents, accessed 20 March 2013.
37. UK Government White Paper, June 2011, 'The Natural Choice: Securing the Value of Nature' (London: The Stationery Office, 2011) Crown Copyright, p. 5.

38. *The Times*, Germany Pledges to Shut All Nuclear Plants by 2022,' 31 May 2011.
39. UK Government White Paper, June 2011, p. 2.
40. Barack Obama, 27 January 2010, State of the Union Address, http://www.whitehouse.gov/the-press-office/remarks-president-state-union-address, accessed 21 March 2013.
41. Barack Obama, 21 February 2013, State of the Union Address, http://www.whitehouse.gov/the-press-office/2013/02/12/remarks-president-state-union-address, accessed 13 February 2013.
42. David Cameron, 11 December 2012, Oral Evidence Taken before the Liaison Committee, transcript located at http://www.publications.parliament.uk/pa/cm201213/cmselect/cmliaisn/uc484-ii/uc48401.htm, accessed 10 January 2014.
43. David Cameron, 11 December 2012.
44. David Cameron, 11 December 2012.
45. Peter Altmaier, 29 January 2013, *Spiegel Online International*, http://www.spiegel.de/international/germany/german-environment-ministry-plans-to-cap-subsidies-for-renewables-a-880301.html, accessed 4 April 2013.
46. Michael Birnbaum, 1 April 2012, *The Washington Post*, 'European Industry Flocks to U.S. for Cheaper Natural Gas,' http://www.washingtonpost.com/world/europe/european-industry-flocks-to-cheap-us-gas/2013/04/01/454d06ea-8-a2c-11e2–98d9–3012c1cd8d1e_story.html, accessed 2 April 2013.
47. Philipp Roesler, quoted in http://www.denverpost.com/digitalfirstmedia/cI_22922268/european-industry-flocks-us-cheaper-natural-gas, accessed 2 April 2013.
48. *The Times*, 11 January 2014, http://www.thetimes.co.uk/tto/environment/article3972094.ece, accessed 13 January 2014.
49. David Cameron, 13 January 2014, https://www.gov.uk/government/news/local-councils-to-receive-millions-in-business-rates-from-shale-gas-developments, accessed 13 January 2014.
50. Barack Obama, 14 November 2012, Press Conference, transcript located at http://www.nytimes.com/2012/11/14/us/politics/running-transcript-of-president-obamas-press-conference.html?pagewanted=10&_r=0, accessed 5 January 2014.
51. Rajendra Pachauri, 22 February 2013, Interview with *The Australian* newspaper, http://www.theaustralian.com.au/news/nothing-off-limits-in-climate-debate/story-e6frg6n6–1226583112134, accessed 21 March 2013.
52. Barack Obama, 14 November 2012, Press Conference.
53. Barack Obama, 14 November 2012, Press Conference.
54. Barack Obama, 21 February 2013, State of the Union Address, http://www.whitehouse.gov/the-press-office/2013/02/12/remarks-president-state-union-address,accessed 20 July 2014.

4 Tyranny, Freedom, Democracy

1. For the American reader, the reference here is to soccer, though similar statements have no doubt been made throughout the history of American football.
2. Bill Shankly, Official Club Website of Liverpool FC, http://www.liverpoolfc.com/news/latest-news/bill-shankly-in-quotes, accessed 1 February 2014.

3. Bill Clinton, 16 July 1992, Address Accepting the Presidential Nomination at the Democratic National Convention in New York, http://www.presidency. ucsb.edu/ws/?pid=25958, accessed 1 February 2014.

4. Tony Blair, *A Journey*, (London: Hutchinson, 2010) p. 229.

5. Winston Churchill, 11 November 1947, Speech to Parliament, Hansard, Vol. 444, cc. 207, http://hansard.millbanksystems.com/commons/1947/nov/11/ parliament-bill#column_207, accessed 2 February 2014.

6. The meaning of terms like liberal and liberalism are contested by protagonists across the political spectrum. For ease of understanding I assume that, while specific meaning is located in the context of use, liberalism broadly understood contains sufficient commitment to equality, rights, political freedoms and representative democracy to be used coherently in this chapter.

7. There is not the scope here to explore fully the dangers of what an apparently liberal majority can do to a minority: consider that it was only in 1967 that homosexuality was legalised in England and Wales.

8. John Adams, *The Works of John Adams, Vol. 6, Defence of the Constitutions Vol. III* (Boston: Charles C. Little and James Brown, 1851) pp. 131–132.

9. Bill Clinton, 24 March 1999, Statement on Kosovo, transcript at http://mill-ercenter.org/president/speeches/detail/3932, accessed 10 December 2013.

10. Bill Clinton, 24 March 1999.

11. Francis Fukuyama, *The End of History and the Last Man* (London and New York: Penguin Books, 1992).

12. More will be said later about the nature of freedom, especially when it is being 'given to' or 'forced upon' others.

13. Tony Blair, 24 April 1999, 'Prime Minister's speech: Doctrine of the International community at the Economic Club, Chicago', http://webar-chive.nationalarchives.gov.uk/20061004085342/http://number10.gov.uk/ page1297, accessed 2 February 2014.

14. Tony Blair, 24 April 1999.

15. George W. Bush, 17 September 2001, President Bush's comments from the Pentagon on Osama bin Laden, transcript at http://www.washingtonpost. com/wp-srv/nation/specials/attacked/transcripts/bush091701.html, accessed 17 December 2013.

16. George W. Bush, 20 September 2001, Transcript of President Bush's Address to a Joint Session of Congress and the Nation, located at http://www.washington-post.com/wp-srv/nation/specials/attacked/transcripts/bushaddress_092001. html, accessed 30 January 2014.

17. George W. Bush, 13 March 2002, Press Conference in the James S. Brady Briefing Room.

18. Tony Blair, Statement in response to terrorist attacks in the United States, 11 September 2001.

19. George W. Bush, 8 November 2001, *Press Conference: Prime Minister Tony Blair and President George Bush*, transcript at http://webarchive.nationalarchives. gov.uk/20040830224728/http://www.pm.gov.uk/output/page1644.asp, accessed 10 December 2013.

20. Tony Blair, 17 March, *Minute to Jonathan Powell*, Evidence to the Iraq Inquiry, http://www.iraqinquiry.org.uk/media/50751/Blair-to-Powell-17March2002-minute.pdf, accessed 17 December 2013.

21. George W. Bush and Tony Blair Joint Press Conference, 6 April 2002, Transcript Located at Http://Georgewbush-Whitehouse.Archives.Gov/News/Releases/2002/04/20020406–3.Html, Accessed 10 February 2014.
22. Blair and Bush, Joint Press Conference, 6 April 2002.
23. Blair and Bush, Joint Press Conference, 6 April 2002.
24. Blair and Bush, Joint Press Conference, 6 April 2002.
25. Tony Blair 16 March 2003, Prime Minister outlines vision for Iraq and the Iraqi people, transcript at http://webarchive.nationalarchives.gov.uk/20040830224728/http://www.pm.gov.uk/output/Page3283.asp, accessed 21 July 2014.
26. Tony Blair, Statement opening Iraq debate in Parliament, 18 March 2003, transcript located at http://webarchive.nationalarchives.gov.uk/+/http:/www.number10.gov.uk/Page3294, accessed 10 February 2014.
27. Blair, Statement opening Iraq debate in Parliament, 18 March 2003.
28. Limited details of the operation were supplied immediately afterwards in a White House press briefing. See http://www.whitehouse.gov/the-press-office/2011/05/02/press-briefing-senior-administration-officials-killing-osama-bin-laden, accessed 10 February 2014.
29. Isaiah Berlin, 1958, *Two Concepts of Liberty*, in Isaiah Berlin, *Four Essays on Liberty* (Oxford: Oxford University Press, 1969).
30. Clearly this subject deserves an entire book of its own but I make the point here to illustrate the ridiculousness of trying to fast-track culturally, socially and religiously distant societies towards some liberal Western 'ideal' that has its own unique history. For an outstanding account of the complexity of one aspect of political history I recommend Jens Bartelson, *A Genealogy of Sovereignty* (Cambridge: Cambridge University Press, 1995).
31. 24 April 2013, 'Kosovo and Serbia Make a Deal', *Foreign Affairs*, http://www.foreignaffairs.com/articles/139346/nikolas-k-gvosdev/kosovo-and-serbia-make-a-deal, accessed 10 December 2013.
32. Republic of Kosovo Ministry of Foreign Affairs, http://www.mfa-ks.net/?page=2,33, accessed 2 February 2014.
33. *The Times*, 16 March 2013, Anthony Loyd, 'Ten Years after Saddam, Iraqi Kurds Have Never Had It So Good'.
34. *The Times*, 16 March 2013.
35. Iraq Body Count, http://www.iraqbodycount.org/database/, accessed 7 February 2014. There are several 'body count' sources, each with its own methodology and each claiming to be the most comprehensive.
36. UN Security Council Statement, 17 December 2013, http://www.un.org/News/Press/docs/2013/sc11218.doc.htm, accessed 21 July 2014.
37. UN News Centre, 20 December 2013. Also, see Transparency International Corruption Perceptions Index, p. 3, http://www.transparency.org/cpi2013/results, accessed 21 July 2014.
38. Tony Blair, 7 April 2002, Speech at the George Bush Senior Presidential Library, transcript at http://webarchive.nationalarchives.gov.uk/20040830224728/http://www.pm.gov.uk/output/Page1712.asp, accessed 7 February 2014.
39. George W. Bush, 1 May 2003, Speech Aboard the USS Abraham Lincoln, transcript located at http://edition.cnn.com/2003/US/05/01/bush.transcript/, accessed 7 February 2014.
40. Blair, 24 April 1999.

41. United Nations Security Council Resolution 1244, 10 June 1999, 'Security Council Welcoming Yugoslavia's Acceptance of Peace Principles, Authorizes Civil, Security Presence in Kosovo', https://www.un.org/News/Press/docs/1999/19990610.SC6686.html, accessed 10 February 2014.
42. Hamid Karzai, 2 February 2014, *The Sunday Times*.
43. *The Guardian*, 13 February 2014.
44. While I opt for the spelling 'Moammar Gaddafi,' I have retained the alternative spelling found in President Obama's speech transcripts: 'Muammar Qaddafi'.
45. Barack Obama, 28 March 2011, 'Remarks by the President in Address to the Nation on Libya', transcript located at http://www.whitehouse.gov/the-press-office/2011/03/28/remarks-president-address-nation-libya, accessed 21 December 2013.
46. Barack Obama, 28 March 2011.
47. United Nations Security Council Resolution 1973, 17 March 2011, http://www.un.org/en/ga/search/view_doc.asp?symbol=S/RES/1973(2011), accessed 10 February 2014.
48. Barack Obama, 28 March 2011.
49. Barack Obama, 28 March 2011.
50. Human Rights Watch World Report 2013, Libya, http://www.hrw.org/world-report/2013/country-chapters/libya, accessed 10 February 2014.
51. Barack Obama, 21 August 2012, http://edition.cnn.com/2012/08/20/world/meast/syria-unrest/, accessed 10 February 2014.
52. Barack Obama, 4 September 2013, http://www.whitehouse.gov/the-press-office/2013/09/04/remarks-president-obama-and-prime-minister-reinfeldt-sweden-joint-press-, accessed 11 February 2014.
53. Vision Critical poll for Express Online, 30 August 2013, http://www.express.co.uk/news/uk/425446/Government-lose-vote-on-war-in-Syria-as-only-EIGHT-per-cent-of-Brits-want-urgent-strikes, accessed 11 February 2014.
54. Philip Hammond, 1 February 2014, Munich Security Conference, reported in The Telegraph, 2 February 2014.
55. Philip Hammond, 1 February 2014.
56. CNN, 30 December 2013, http://edition.cnn.com/video/?/video/politics/2013/12/30/tsr-dnt-acosta-afghanistan-new-low.cnn&iref=allsearch&video_referrer=http per cent3A per cent2F per cent2Fedition.cnn.com per cent2Fsearch per cent2F per cent3Fquery per cent3Dafghanistan per cent2Bwar per cent26primaryType per cent3Dmixed per cent26sortBy per cent3Drelevance per cent26intl per cent3Dtrue per cent26x per cent3D0 per cent26y per cent3D0, accessed 2 February 2014.
57. Transatlantic Trends 2013 Survey, http://trends.gmfus.org/files/2013/09/TTrends-2013-Key-Findings-Report.pdf, accessed 2 February 2014.
58. *The Times*, 2 February 2014.

5 Gendering Military Intervention

1. Tony Blair, 24 April 1999, Chicago Speech.
2. The UN's 2005 World Summit Outcome outlined a new 'Responsibility to Protect' doctrine (http://unpan1.un.org/intradoc/groups/public/documents/un/unpan021752.pdf. However, it remains highly contentious and appears

to have had little or no impact on subsequent proposed or actual military interventions.

3. Jean Bethke Elshtain, 'Women and War: Ten Years On', *Review of International Studies*, Vol. 24 (1998) p. 453.
4. Joshua S. Goldstein, *War and Gender: How Gender Shapes the War System and Vice Versa* (Cambridge: Cambridge University Press, 2001) p. 1.
5. Tony Blair, 26 April 1999, http://webarchive.nationalarchives.gov.uk/20040621031906/http://number10.gov.uk/page1300, accessed 21 July 2014.
6. Tony Blair, 25 September 2001, http://webarchive.nationalarchives.gov.uk/20040621031906/http://number10.gov.uk/page1604, accessed 21 July 2014.
7. Tony Blair, 18 March 2003, http://webarchive.nationalarchives.gov.uk/20040621031906/http://number10.gov.uk/page3294, accessed 14 January 2014.
8. George W. Bush, 21 September 2004, Speech to the UN, http://www.theguardian.com/world/2004/sep/21/iraq.usa3, accessed 14 January 2014.
9. *Mail on Sunday*, 31 March 2013; Channel 4, 5.30pm News, 31 March 2013.
10. Krista Hunt, '"Embedded Feminism" and the War on Terror', in Krista Hunt and Kim Rygiel (eds), *(En) Gendering the War on Terror: War Stories and Camouflaged Politics* (Aldershot: Ashgate, 2006) p. 51.
11. While there is not the scope here for a full exploration of the arguments involved, further reading is highly recommended. A good starting point is Krista Hunt and Kim Rygiel (eds), *(En)Gendering the War on Terror: War Stories and Camouflaged Politics* (Aldershot: Ashgate, 2006).
12. Jean Bethke Elshtain, *Just War against Terror* (New York: Basic Books, 2004) pp. 39–40.
13. BBC, 14 January 2012, 'What Future for Afghan Woman Jailed for Being Raped?http://www.bbc.co.uk/news/world-south-asia-16543036, accessed 10 January 2014.
14. I acknowledge that men can also be trapped in abusive relationships.
15. http://www.huffingtonpost.com/2013/08/05/cleveland-courage-fund_n_3707585.html, accessed 13 January 2014.
16. BBC, 14 January 2012, 'What Future for Afghan Woman Jailed for Being Raped?http://www.bbc.co.uk/news/world-south-asia-16543036, accessed 10 January 2014.
17. For an extensive analysis of Augustine's writings on just war and the place of the subject therein, see Peter Lee, *A Genealogy of the Ethical Subject in the Just War Tradition*, PhD Thesis, accepted by King's College London, 1 August 2010.
18. Augustine, *The City of God against the Pagans* (Cambridge: Cambridge University Press, 1998) Book XXII Chapter 6, p. 1118.
19. Augustine, *Questions on the Heptateuch*, Book VI, Chapter 10, in Gregory M. Reichberg, Henrik Syse, and Endre Begby (eds) *The Ethics of War* (Oxford: Blackwell Publishing, 2006) p. 82.
20. Augustine, *Against Faustus the Manichean*, Ch. XXII.74, in E. L. Fortin and D. Kries (eds), *Augustine: Political Writings*, Trans. M. W. Tkacz and D. Kries (Indianapolis: Hackett, 1994) pp. 221–222.
21. See the Charter of the United Nations, Articles 51 and 39, https://www.un.org/en/documents/charter/chapter7.shtml, accessed 15 January 2014.

22. George W. Bush, 16 September 2001, Remarks from the White House, http://georgewbush-whitehouse.archives.gov/news/releases/2001/09/20010916-2.html, accessed 15 January 2014.

23. Matthew Ch. 12, vs. 30, New International Version.

24. Nabil Shaath, quoting George W. Bush, BBC documentary promotion 6 October 2005, *Elusive Peace: Israel and the Arabs,* http://www.bbc.co.uk/pressoffice/press-releases/stories/2005/10_october/06/bush.shtml, accessed 16 January 2014.

25. Blair, Interview with Michael Parkinson, 4 March 2006, transcript located at http://news.bbc.co.uk/1/hi/uk_politics/4773874.stm, accessed 16 January 2014.

26. Elshtain, *Just War against Terror,* p. 38ff.

27. Elshtain, *Just War Against Terror,* p. 241.

28. Jean Bethke Elshtain, *Sovereignty: God, State, and Self* (New York: Basic Books, 2008) p. 241.

29. Tony Blair, 19 October 2001, Press Conference, http://webarchive.nationalarchives.gov.uk/20040621031906/http://number10.gov.uk/page1634, accessed 21 July 2014.

30. George W. Bush, 21 September 2004, Speech to the UN, http://www.theguardian.com/world/2004/sep/21/iraq.usa3, accessed 14 January 2014.

31. Human Rights Watch, 2010, *The 'Ten-Dollar Talib' and Women's Rights: Afghan Women and the Risks of Reintegration and Reconciliation,* p. 45. Report located at http://www.hrw.org, accessed 9 March 2012.

32. *Time Magazine,* 9 August 2010. Image and further details of Aisha's story located at http://content.time.com/time/magazine/article/0,9171,2007415,00.html, accessed 16 January 2014.

33. Aisha added further detail to the original 2010 story after facial reconstruction in the US, 26 February 2013, http://www.nydailynews.com/news/national/mutilated-afghan-wife-debuts-new-nose-article-1.1273781, accessed 16 January 2014.

34. Declaration from Afghanistan's Ulema Council, 2 March 2012, Section 5, Para. F.1.D. The original can be located at the website of President Karzai, http://president.gov.af/fa/news/7489. Translation located at http://afghanistananalysis.wordpress.com/2012/03/04/english-translation-of-ulema-councils-declaration-about-women/, accessed 16 January 2014.

35. Galatians 3:28, New International Version.

36. In some Protestant denominations there is equality in leadership positions – Methodists, Presbyterians, Salvationists – but the majority of Pentecostal churches share very similar views on women and leadership to the Catholic and Orthodox churches.

37. *The Times,* 17 January 2014.

38. Paul Rodgers, *United States Constitutional Law: An Introduction* (Jefferson, North Carolina: McFarland, 2011) p. 109.

39. RCM, RCN, RCOG, Equality Now, UNITE, *Tackling FGM in the UK: Intercollegiate Recommendations for Identifying, Recording, and Reporting* (London: Royal College of Midwives, 2013) p. 6.

40. W. K. Jones, J. Smith, B. Kieke Jr and L. Wilcox, 'Female Genital Mutilation. Female Circumcision. Who Is at Risk in the U.S.?' *Public Health Reports,* Vol. 112, No. 5 (September–October 1997) pp. 368–377.

41. The United Nations Convention on the Rights of the Child, 2 September 1990, Article 37a, p. 10, http://www.unicef.org.uk/Documents/Publication-pdfs/UNCRC_PRESS200910web.pdf, accessed 16 January 2014.
42. George W. Bush, 1 May 2003, Speech on USS Abraham Lincoln, Http://News.Bbc.Co.Uk/1/Hi/World/Americas/2994345.Stm, Accessed 20 January 2014.
43. Barack Obama, 28 March 2011, Address to the Nation on Libya, http://www.whitehouse.gov/the-press-office/2011/03/28/remarks-president-address-nation-libya, accessed 14 January 2014.
44. Barack Obama, 24 September 2013, Speech to the United Nations on Syria, Iran and war, transcript at http://www.washingtonpost.com/blogs/worldviews/wp/2013/09/24/read-full-text-of-obamas-speech-to-the-united-nations-on-syria-iran-and-war/, accessed 14 January 2014.
45. David Cameron, 2 October 2013, Speech to Conservative Party Conference, transcript at http://www.telegraph.co.uk/news/politics/david-cameron/10349712/Conservative-Party-Conference-David-Camerons-speech-in-full.html, accessed 20 January 2014.
46. Bruce Tucker, 'Lynndie England, Abu Ghraib, and the New Imperialism', *Canadian Review of American Studies*, Vol. 39, No. 1 (2008) pp. 83–100.
47. Tony Blair, 8 October 2001, http://webarchive.nationalarchives.gov.uk/20040621031906/http://number10.gov.uk/page1621, accessed 21 July 2014.
48. Tony Blair, 14 November 2001, http://webarchive.nationalarchives.gov.uk/20040621031906/http://number10.gov.uk/page1668, accessed 21 July 2014.
49. Lisa K. Foster and Scott Vince, *California's Women Veterans: The Challenges and Needs of Those Who Served* (California, California Research Bureau, 2009) p. 34.
50. United States Department of Defense, Sexual Assault Prevention and Response Office, http://www.sapr.mil/, accessed 2 February 2014.
51. Madeleine Moon, 1 February 2014, *The Observer*.

6 Drone Wars

1. The word 'drone' is contested as a descriptor. Governments and air forces prefer terms like Unmanned Aerial Vehicles or Remotely Piloted Aircraft, while the International Civil Aviation Organisation (ICAO) opts for Remotely Piloted Aircraft System (RPAS). 'Drone' is preferred by opposition groups because it connotes a more sinister degree of dehumanized, detached activity. As a result of the lack of public engagement by the US, UK and other governments at the time when Predator and Reaper were being developed and deployed, the word 'drone' has become ubiquitous in public discourse and will be used throughout this chapter in conjunction with Predator and Reaper, the two armed drone types currently deployed by the US and UK.
2. http://www.youtube.com/watch?v=GSIpatLqSYc, accessed 23 January 2014.
3. There are a number of indigenous anti-government groups operation in Afghanistan, not only the mainly Pashtun Taliban. However, for simplification, those indigenous anti-government, anti-NATO Afghan fighters will be referred to here under the rubric 'Taliban', while non-Afghan fighters will be identified with Al-Qaeda.

4. The Israeli Air Force is the only other major air force currently carrying out strike operations from drones. This discussion, however, will focus on the UK and US.

5. Barack Obama, 23 May 2013, National Security Address, http://www.white-house.gov/the-press-office/2013/05/23/remarks-president-national-defense-university, accessed 24 January 2014.

6. *The Guardian*, 5 July 2011, p. 1, referring to an incident on 25 March 11, located at http://www.guardian.co.uk/uk/2011/jul/05/afghanistan-raf-drone-civilian-deaths, accessed 24 January 2014.

7. International Human Rights and Conflict Resolution Clinic of Stanford Law School (Stanford Clinic) and the Global Justice Clinic at New York University School of Law (NYU Clinic), Report: *Living Under Drones: Death, Injury and Trauma to Civilians from US Drone Practices in Pakistan*, report located at http://livingunderdrones.org/report/, accessed 7 November 2012. The report relies on statistics from the Bureau of Investigative Journalism which state: between June 2004 and September 2012, between 2,562 and 3,325 people were killed by drone strikes in Pakistan, with between 474 and 881 of those being civilians. By January 2014 the claimed upper figures had risen to 3,646 killed, including 951 civilians. See also The Bureau of Investigative Journalism, 24 January 2014, http://www.thebureauinvestigates.com/category/projects/drones/drones-pakistan/, accessed 24 January 2014.

8. Rwanda Genocide and the United Nations, http://www.un.org/en/prevent-genocide/rwanda/education/rwandagenocide.shtml, accessed 24 January 2014.

9. In a British poll up to 67 per cent of respondents supported drone strikes if no civilian casualties would result, http://www.huffingtonpost.co.uk/2013/03/25/drone-strikes-uk-killings_n_2951412.html, accessed 24 January 2014. Drone strikes are far less controversial in the US where polls consistently show overwhelming and more permissive support for their use against suspected terrorists, http://nbcpolitics.nbcnews.com/_news/2013/06/05/18780381-poll-finds-overwhelming-support-for-drone-strikes?lite, accessed 24 January 2014.

10. In a proliferating literature, the following diverse examples illustrate a common perception: Kristen Boon (ed.) 'Terrorism: The Changing Nature of War', *Commentary on Security Documents*, Vol. 127 (Oxford: Oxford University Press, 2012); 'Drone Technology Raises Questions on Evolving Nature of War and Its Conduct', 5 November 2012, *National Catholic Enquirer*, http://ncronline.org/news/politics/'irresistible-attractions'-drones, accessed 25 June 2013; 'How the Predator Drone Changed the Character of War', November 2013, http://www.smithsonianmag.com/history/how-the-predator-drone-changed-the-character-of-war-3794671/, accessed 12 December 2013.

11. Bureau of Investigative Journalism, 26 January 2014, http://www.thebureauinvestigates.com/category/projects/drones/drones-pakistan/, accessed 26 January 2014.

12. *The Times*, 27 November 2011, http://www.thebureauinvestigates.com/category/projects/drones/drones-pakistan/, accessed 12 December 2014.

13. The International Security Assistance Force comprises US, UK, and other NATO members, Afghan security personnel, and contributions from additional countries around the world. For details see http://www.isaf.nato.int/troop-numbers-and-contributions/index.php, accessed 27 April 2014.

14. UN Security Council Report, 13 June 2013, 'The Situation in Afghanistan and Its Implications forInternational Peace and Security', http://www.securitycouncilreport.org/atf/cf/ per cent7B65BFCF9B-6D27–4E9C-8CD3-CF6-E4FF96FF9 per cent7D/s_2013_350.pdf, accessed 27 January 2014.

15. UN Human Rights Council, 9 April 2013, 'Report of the Special Rapporteur on Extrajudicial, Summary or Arbitrary Executions, Christof Heyns', http://www.ohchr.org/Documents/HRBodies/HRCouncil/RegularSession/Session23/A-HRC-23–47_En.pdf, accessed 27 January 2014.

16. *Reuters*, 10 March 2013, http://www.reuters.com/article/2013/03/10/us-usa-afghanistan-drones-idUSBRE92903520130310, accessed 11 November 2013.

17. Barack Obama, 23 May 2013, National Security Address, http://www.whitehouse.gov/the-press-office/2013/05/23/remarks-president-national-defense-university, accessed 24 January 2014.

18. Barack Obama, 23 May 2013, National Security Address.

19. *The Guardian*, 5 July 2011, p. 1, referring to an incident on 25 March 2011, located at http://www.guardian.co.uk/uk/2011/jul/05/afghanistan-raf-drone-civilian-deaths, accessed 24 January 2014.

20. Drone Wars UK, located at http://dronewars.net/.

21. See Note 7 for details.

22. Supreme Court Judgement, 19 June 2013, 'Smith, Ellis, Allbutt and Others *v* The Ministry of Defence', [2013] UKSC 41, p. 21, para. 57, http://www.supremecourt.gov.uk/decided-cases/docs/UKSC_2012_0249_Judgment.pdf, accessed 26 June 2013.

23. George Bush, 20 September 2001, 'President Bush's Address to a Joint Session of Congress and the Nation', transcript located at http://www.washingtonpost.com/wp-srv/nation/specials/attacked/transcripts/bushaddress_092001.html, accessed 30 November 2013.

24. UN Charter, 26 June 1945, Article 51, http://www.un.org/en/documents/charter/chapter7.shtml, accessed 20 October 2013.

25. I refer here to the Western just war tradition whose philosophical roots can be traced from the classical Greek period, through Augustine and subsequent Christian scholars to the Enlightenment challenges to divine order, and on to the present.

26. Barack Obama, 23 May 2013, National Security Address, http://www.whitehouse.gov/the-press-office/2013/05/23/remarks-president-national-defense-university, accessed 24 January 2014.

27. For further examples of the analysis and application of just war criteria in a vast field see Michael Walzer, *Just and Unjust Wars*, 4th Edition (New York: Basic Books, 2006); Richard Norman, *Ethics, Killing and War* (Cambridge: Cambridge University Press, 1995) p. 118; Alex Bellamy, *Just Wars: From Cicero to Iraq* (Cambridge: Polity Press, 2006) p. 121–123; and Nicholas Rengger, 'The Ethics of War: The Just War Tradition', in D. Bell (ed.), *Ethics and World Politics* (Oxford: Oxford University Press, 2010) pp. 296–298.

28. A separate debate is brewing over surveillance drones intruding upon individual freedoms and the right to privacy – at least in countries where there are individual freedoms. For now, while drones are used as weapon delivery systems the emphasis remains with the lethal application of force.

29. Pilots fly the drones; sensor operators carry out the surveillance tasks and control the missiles onto their targets.

30. Chris Cole, Mary Dobbing and Amy Hailwood, *Convenient Killing: Armed Drones and the 'Playstation' Mentality* (Fellowship of Reconciliation: Oxford, 2010).
31. Peter W. Singer, *Wired for War* (New York: Penguin, 2009) p. 332.
32. Peter Olsthoorn, *Military Ethics and Virtues: An Interdisciplinary Approach for the 21st Century* (New York: Routledge, 2011) p. 126.
33. Medea Benjamin, *Drone Warfare: Killing by Remote Control* (New York and London: OR Books, 2012).
34. For an extended exploration of this point see Peter Lee, 'Rights, Wrongs and Drones: Remote Warfare, Ethics and the Challenge of Just War Reasoning', *Air Power Review*, Vol. 16, No. 3 (Autumn/Winter 2013) pp. 30–49.
35. Personal communication with author, 16 July 2013, Creech Air Force Base.
36. W. Chappelle, K. McDonald and K. McMillan, *Important and Critical Psychological Attributes of USAF MQ-1 Predator and MQ-9 Reaper Pilots According to Subject Matter Experts* (Wright-Patterson AFB, OH: Air Force Research Laboratory, 2011); J. L. Otto and Capt. B. J. Webber, 'Mental Health Diagnoses and Counselling Among Pilots of Remotely Piloted Aircraft in the United States Air Force', available from http://timemilitary.files.wordpress.com/2013/04/pages-from-pages-from-msmr_mar_2013_External_causes_of_tbi.pdf accessed 25 January 2014.
37. ISR: Intelligence, surveillance and reconnaissance.
38. IED: Improvised Explosive Device.
39. ROE: Rules of Engagement.
40. Personal communication from a UK Reaper pilot, 18 August 2013, individual based at Creech Air Force Base.
41. Bureau of Investigative Journalism, 6 January 2014, http://www.thebureauinvestigates.com/2014/01/06/a-changing-drone-campaign-us-covert-actions-in-2013/, accessed 12 January 2014.
42. Protocol Additional to the Geneva Conventions of 12 August 1949, and relating to the Protection of Victims of International Armed Conflicts (Protocol I), 8 June 1977, Article 51.5.b, http://www.icrc.org/applic/ihl/ihl.nsf/Article.xsp?action=openDocument&documentId=4BEBD9920AE0AEAEC12563CD0051DC9E, accessed 20 January 2014.

7   It's All Your Fault

1. Ben Bernanke, 28 November 2006, Speech to the National Italian American Foundation, New York, transcript at http://www.federalreserve.gov/newsevents/speech/bernanke20061128a.htm, accessed 10 December 2013.
2. Gordon Brown, 22 March 2006, Budget Speech, Hansard, Column 288, located at http://www.publications.parliament.uk/pa/cm200506/cmhansrd/vo060322/debtext/60322–05.htm accessed 24 January 2014.
3. Gordon Brown 21 March 2007, Budget Speech, located at http://webarchive.nationalarchives.gov.uk/20100407010852/http://www.hm-treasury.gov.uk/bud_budget07_speech.htm, accessed 24 January 2014.
4. Queen Elizabeth, 6 November 2008, *The Daily Mail*.
5. Alan Greenspan, 23 October 2008, Testimony to the House Committee of Government Oversight and Reform, transcript at http://blogs.wsj.com/

economics/2008/10/23/greenspan-testimony-on-sources-of-financial-crisis/, accessed 10 January 2014.

6. The Royal Swedish Academy of Sciences, 14 October 1997, http://www. nobelprize.org/nobel_prizes/economic-sciences/laureates/1997/press.html, accessed 26 February 2014.

7. Charles Bean, 25 August 2009, *The Great Moderation, the Great Panic and the Great Contraction*, Schumpeter Lecture, Annual Congress of the European Economic Association, Barcelona, transcript.http://www.bankofengland. co.uk/archive/Documents/historicpubs/speeches/2009/speech399.pdf, accessed 25 February 2014.

8. Bean, *The Great Moderation*, p. 2.

9. Bean, *The Great Moderation*.

10. Nouriel Roubani, 7 September 2006, International Monetary Fund seminar, Washington DC, transcript at http://www.economonitor.com/ nouriel/2010/09/02/economonitor-flashback-roubinis-imf-speech-september-7-2006/, accessed 26 February 2014.

11. Nouriel Roubani, 7 September 2006.

12. Nouriel Roubani, 7 September 2006.

13. *New York Times*, 15 August 2008, http://www.nytimes.com/2008/08/17/ magazine/17pessimist-t.html?_r=0, accessed 26 February 2008.

14. Alan Greenspan, 23 October 2008, Testimony to the House Committee of Government Oversight and Reform, transcript at http://blogs.wsj.com/ economics/2008/10/23/greenspan-testimony-on-sources-of-financial-crisis/, accessed 10 January 2014.

15. Financial Crisis Inquiry Commission, *The Financial Crisis Inquiry Report: Final Report of the National Commission on the Causes of the Financial and Economic Crisis in the United States* (New York: Public Affairs, 2011).

16. The Commission members and the party affiliation of their Democrat and Republican nominators are as follows: Phil Angelides (chairman) (Dem); Bill Thomas (vice chairman) (Rep); Brooksley Born (Dem); Byron Georgiou (Dem); Bob Graham (Dem); Keith Hennessey (Rep); Douglas Holtz-Eakin (Rep); Heather Murren (Dem); John W. Thompson (Dem); Peter J. Wallison (Rep).

17. *The Financial Crisis Inquiry Report*, 2011, pp. xvii-xxv.

18. *The Financial Crisis Inquiry Report*, 2011, p. 444.

19. For a rebuttal of Wallison's claims see David Min, 12 July 2011, 'Why Wallison Is Wrong About the Genesis of the U.S. Housing Crisis', http://www.americanprogress.org/issues/housing/report/2011/07/12/10011/why-wallison-is-wrong-about-the-genesis-of-the-u-s-housing-crisis/, accessed 28 February 2014.

20. Howard Davies, *The Financial Crisis: Who is to Blame?* (Cambridge: Polity Press, 2010) p. 3.

21. *The Financial Crisis Inquiry Report*, 2011, p. 448.

22. Davies, *The Financial Crisis*.

23. Davies, *The Financial Crisis*, p. 213.

24. Andrew Hindmoor and Allan McConnell, 'Why Didn't They See It Coming? Warning Signs, Acceptable Risks and the Global Financial Crisis', *Political Studies*, Vol. 61 (2013) pp. 543–560. For further reading among an extensive and growing literature see also Erik Banks, *See No Evil: Uncovering the*

Truth Behind the Financial Crisis (Basingstoke: Palgrave Macmillan, 2011); Christopher Hood, *The Blame Game: Spin, Bureaucracy, and Self-Preservation in Government* (Princeton NJ: Princeton University Press, 2011).

25. It would take a book-length genealogy to unravel some of the interconnections between these various strands of thought on how the political leader is formed and self-forming as a subject of contemporary political discourse. The purpose of this short section is to demonstrate the complexity of the financial truth wars and any attempt to apportion blame in a meaningful way.

26. This sentence will prove controversial or offensive to some. I merely point to the Christian televangelists who have been imprisoned for fraud, paedophile priests, and the recent rise in religiously-motivated suicide bombers to highlight that faith claims and personal behaviour are not always, and are perhaps rarely, consistent.

27. Mervyn King, 26 February 2009, Governor of the Bank of England, Appearance before the Treasury Select Committee, Q2355, http://www.publications.parliament.uk/pa/cm200809/cmselect/cmtreasy/144/09022603.htm, accessed 3 March 2014.

28. Despite an extensive search I could not locate the origin of the quote, though it has taken on a life of its own in the blogosphere.

29. Ben S. Bernanke, 14 April 2009, Speech: Four Questions About the Financial Crisis, transcript located at http://www.federalreserve.gov/newsevents/speech/bernanke20090414a.htm, accessed 4 March 2014.

30. *The Financial Crisis Inquiry Report*, p. 448.

31. Robert J. Schiller, *The Subprime Solution: How Today's Global Financial Crisis Happened, and What To Do About It* (Princeton: Princeton University Press, 2008) p. 6.

32. Statistics up to 27 February 2014 are available from the Department for Communities and Local Government, https://www.gov.uk/government/statistical-data-sets/live-tables-on-social-housing-sales, accessed 6 March 2014.

33. While the origin of the phrase might be disputed, the UK can lay claim to a greater number of castles than the US. Given the historical, socio-political implications of having a high number of castles, I do not see this as a British boast.

34. Details at https://www.gov.uk/right-to-buy-buying-your-council-home/discounts, accessed 7 March 2014. Cash limits on the discounts have applied at different times. In 2014 the maximum value of the discount is £75,000 across England or £100,000 in London. The British prefer the term 'flats' to 'apartments'.

35. Scott Stern, CEO of Lenders One, 10 April 2008, Testimony before the Senate Banking Committee Washington, transcript at http://www.gpo.gov/fdsys/pkg/CHRG-110shrg50396/html/CHRG-110shrg50396.htm, accessed 4 March 2014.

8   Governing Greed

1. *The Sunday Times*, 18 December 2005, 'Bonus Bonanza as Bankers Strike Gold'.

2. *New York Times*, 8 November 2005, 'Optimism on Wall Street Over Size of Bonuses', http://www.nytimes.com/2005/11/08/business/08place.html, accessed 10 March 2014.

3. Barack Obama quoted in *Politico*, 3 April 2009, http://www.politico.com/news/stories/0409/20871.html, accessed 10 March 2014.

4. Tony Blair, 23 July 2012, *The Telegraph*, http://www.telegraph.co.uk/news/politics/tony-blair/9422096/Tony-Blair-hanging-bankers-wont-help.html, accessed 10 March 2014.

5. *The Times*, 12 March 2014, 'Pinstripes Are Out But The Good Times Are Back'.

6. *International Business Times*, 16 January 2014, 'Goldman Sachs Pay and Bonuses Hit $12.61bn in 2013', http://www.ibtimes.co.uk/goldman-sachs-pay-bonuses-hit-12–61bn-2013–1432575, accessed 11 March 2014.

7. Though without the effects of inflation taken into account.

8. Exact dates for the crash are disputed depending on the country and criteria used. The years 2007 and 2008 are most commonly seen as the height of the financial catastrophe.

9. For a succinct summary of the major events in the deregulation process see Fiona Tregenna, 'The Fat Years: The Structure and Profitability of the US Banking Sector in the Pre-Crisis Period', *Cambridge Journal of Economics*, Vol. 33 (2009) pp. 609–632.

10. Raghuram G. Rajan, 8 June 2006, Address at the Bank of Spain Conference on Central Banks in the 21st Century, transcript at http://www.imf.org/external/np/speeches/2006/060806.htm, accessed 10 March 2014.

11. For a pre-crash discussion of the relationship between derivatives and risk see Norvald Instefjord, 'Risk and Hedging: Do Credit Derivatives Increase Bank Risk', Research paper for the Bank of England and Banque de France, 14 August 2000, http://www.banque-france.fr/fondation/gb/telechar/papers_d/hedging.pdf, accessed 10 March 2014. Instefjord concluded that credit derivatives could destabilize banks through increased exposure to risk.

12. Financial Crisis Inquiry Commission, *The Financial Crisis Inquiry Report* (New York: Public Affairs, 2011) p. 48.

13. Hyman P. Minsky, *Stabilizing an Unstable Economy* (New Haven, Yale University Press, 1986).

14. *The Financial Crisis Enquiry Report*, p. 48.

15. *The Financial Crisis Enquiry Report*, p. 236.

16. *The Financial Crisis Enquiry Report*, p. 236.

17. Treasury Committee of the UK House of Commons, 10 February 2009, *Banking Crisis: Former Bank Executives*, transcript at: http://www.publications.parliament.uk/pa/cm200809/cmselect/cmtreasy/uc144_vii/uc14402.htm, accessed 12 March 2014.

18. Owen Hargie, Karyn Stapleton and Dennis Tourish, 'Interpretations of CEO Public Apologies for the Banking Crisis: Attributions of Blame and Avoidance of Responsibility', *Organization*, Vol. 17, No. 6 (November 2010) pp. 721–742.

19. Treasury Committee of the UK House of Commons, 10 February 2009, Q. 878.

20. Treasury Committee of the UK House of Commons, 10 February 2009, Q. 883.

21. Treasury Committee of the UK House of Commons, 10 February 2009, Q. 883–884.

22. Treasury Committee of the UK House of Commons, 10 February 2009, Q. 933.

23. Mark Carlson, 'A Brief History of the 1987 Stock Market Crash with a Discussion of the Federal Reserve Response', The Federal Reserve, November 2006, http://www.federalreserve.gov/pubs/feds/2007/200713/200713pap.pdf, accessed 14 March 2014.

24. Christopher J. Neely, 'The Federal Reserve Responds to Crises: September 11th Was Not the First', *Federal Reserve Bank of St. Louis Review*, Vol. 86(2) (March/April 2004) p. 35.
25. Neely, 'The Federal Reserve Responds to Crises, p. 35.
26. Jeffrey M. Lacker, 'Payment System Disruptions and the Federal Reserve Following September 11, 2001', Working Paper 03–16, Federal Reserve Bank of Richmond, 23 December 2003, https://www.richmondfed.org/publications/research/working_papers/2003/pdf/wp03–16.pdf, accessed 15 March 2014.
27. Neely, 'The Federal Reserve Responds to Crises', pp. 12–13.
28. *The Financial Crisis Enquiry Report*, p. 343.
29. Adair Turner, March 2009, '*The Turner Review: A Regulatory Response to the Global Banking Crisis*', The Financial Services Authority, Publication Ref 003289, p. 80, Crown Copyright.
30. Barack Obama, 24 February 2009, Address to Joint Session of Congress, transcript at http://www.whitehouse.gov/the_press_office/Remarks-of-President-Barack-Obama-Address-to-Joint-Session-of-Congress, accessed 12 February 2014.
31. The shadow banking system is generally characterised by three things: first, it does not accept traditional bank deposits; second, and as a consequence of the first point, the institutions are not subject to regulatory oversight in the same way that other banks are; and third, because the institutions are unregulated their activities are not scrutinised.
32. *The Financial Crisis Enquiry Report*, p. 255.
33. Alan Greenspan, 7 April 2010, Written Testimony to the Financial Crisis Inquiry Commission, p. 13, http://fcic-static.law.stanford.edu/cdn_media/fcic-testimony/2010–0407-Greenspan.pdf, accessed 17 March 2014.
34. *The Financial Crisis Enquiry Report*, p. 94.
35. Barack Obama, 14 September 2009, Remarks by the President on Financial Rescue and Reform, Federal Hall, New York, transcript at http://www.whitehouse.gov/the_press_office/Remarks-by-the-President-on-Financial-Rescue-and-Reform-at-Federal-Hall, accessed 17 March 2014.
36. Christopher Bones, *The Cult of the Leader: A Manifesto For More Authentic Business* (Chichester: John Wiley & Sons Ltd, 2011) p. 239.
37. David Cameron, 17 December 2010, EU Summit Press Conference, located at http://www.number10.gov.uk/news/eu-summit-press-conference/, accessed 25 March 2012.
38. Barack Obama, 14 September 2009, Remarks by the President on Financial Rescue and Reform, Federal Hall, New York, transcript at http://www.whitehouse.gov/the_press_office/Remarks-by-the-President-on-Financial-Rescue-and-Reform-at-Federal-Hall, accessed 17 March 2014.
39. Barack Obama, 27 January 2010, State of the Union Address, transcript at http://www.whitehouse.gov/the-press-office/remarks-president-state-union-address, accessed 21 March 2013.
40. David Cameron, 11 December 2012, Oral Evidence Taken before the Liaison Committee, Q. 14, transcript located at http://www.publications.parliament.uk/pa/cm201213/cmselect/cmliaisn/uc484-ii/uc48401.htm, accessed 16 March 2013.
41. *The Times*, 10 February 2014.

42. Simon Lewis, 21 February 2011, *The Telegraph*, http://www.telegraph.co.uk/finance/comment/8336840/Message-to-Brussels-regulators-if-its-not-broken-it-does-not-need-fixing.html, accessed 15 March 2014.
43. Simon Lewis, 21 February 2011.
44. Barack Obama, 14 September 2009, Remarks by the President on Financial Rescue and Reform, Federal Hall, New York, transcript at http://www.whitehouse.gov/the_press_office/Remarks-by-the-President-on-Financial-Rescue-and-Reform-at-Federal-Hall, accessed 17 March 2014.
45. Nouriel Roubini and Stephen Minh, *Crisis Economics: A Crash Course in the Future of Finance* (London and New York: Penguin Books, 2011) pp. 183ff.
46. Ben Bernanke, 3 January 2014, Speech at the Annual Meeting of the American Economic Association, Philadelphia, Pennsylvania, transcript at http://www.federalreserve.gov/newsevents/speech/bernanke20140103a.htm, accessed 19 March 2014.
47. Mark Carney, 18 March 2014, 'One Mission. One Bank. Promoting the Good of the People of the United Kingdom', Mais Lecture at Cass Business School, City University, London, transcript at http://www.bankofengland.co.uk/publications/Documents/speeches/2014/speech715.pdf, accessed 19 March 2014.
48. Carney, 18 March 2014.

9   Who Mentioned the War?

1. Michel Foucault, 'The Ethics of the Concern for Self as a Practice of Freedom', in Michel Foucault and Paul Rabinow (eds) *The Essential Works of Michel Foucault 1954 – 1984 Volume 1: Ethics – Subjectivity and Truth* (New York: The New Press, 1997) p. 290.
2. Foucault, 'The Ethics of the Concern for Self as a Practice of Freedom', p. 290.
3. Foucault preferred the term 'subject' to 'individual', which included within its meaning personal aspects like identity, behaviour and values. For accessibility I opt for the latter terms but my reasoning is informed by Foucault's theorising.
4. G. Federico Mancini, 'Europe: The Case for Statehood', *European Law Journal*, 4(1) (March 1998) p. 30 (my emphasis).
5. The Treaty of Rome, 25 March 1957, http://ec.europa.eu/economy_finance/emu_history/documents/treaties/rometreaty2.pdf, accessed 5 June 2013.
6. Mancini, 'Europe', p. 30.
7. The Maastricht Treaty on European Union came into effect in 1993, the European Union subsequently replacing what was previously referred to as the European Community.
8. Mancini, 'Europe', p. 30.
9. William Hague, 31 May 2013, Speech at the 63rd Königswinter Conference, https://www.gov.uk/government/speeches/britain-and-germany-partners-in-reform, accessed 3 June 2013.
10. For a detailed account of events and the British response see Richard Davis, 'The "Problem of de Gaulle": British Reactions to General de Gaulle's Veto of the UK Application to Join the Common Market', *Journal of Contemporary History*, Vol. 32, No. 4 (1997) pp. 453–464.

11. Jane O'Mahoney, 'Ireland's EU Referendum Experience', *Irish Political Studies*, Vol. 24, No. 4 (2009) p. 430.
12. Hague, 31 May 2013.
13. David Cameron, 23 January 2013, Speech: 'UK and the EU', transcript located at http://www.bbc.co.uk/news/uk-politics-21160684, accessed 4 June 2013.
14. David Cameron, 23 January 2013.
15. IMF Country Report No. 13/156, June 2013, 'Greece: Ex Post Evaluation of Exceptional Access Under the 2010 Stand-By Arrangement', p. 28, http://www.imf.org/external/pubs/ft/scr/2013/cr13156.pdf, accessed 8 June 2013, ©International Monetary Fund 2013.
16. IMF Country Report No. 13/156, June 2013.
17. IMF Country Report No. 13/156, June 2013, p. 27.
18. Olli Rehn, 7 June 2013, *Wall Street Journal*, http://online.wsj.com/article/S B10001424127887323844804578530782011779680.html, accessed 8 June 2013.
19. A particularly effective historical analysis of intrinsic moral order can be found in William E. Connolly, *The Augustinian Imperative* (2nd edition) (Oxford: Rowman & Littlefield Publishers, 2002).
20. *The Times*, 10 February 2014.
21. *The Times*, 18 February 2014.
22. *The Times*, 18 February 2014.
23. Christine Lagarde, 8 March 2013, *Wall Street Journal*, http://wsj.com/article/SB10 0014241278873236288045783480141933 67002.html, accessed 15 June 2013.
24. Jasek Rostowski, 14 September 2011, http://www.novinite.com/view_news. php?id=132086, accessed 9 June 2013.
25. Augustine, *City of God*, Trans. Henry Bettenson (London: Penguin Classics, 2003).
26. John Maynard Keynes cited by James K. Galbraith, 6 December 2012, 'Change of Direction', IG Metall Conference, Berlin, Germany, transcript located at http://utip.gov.utexas.edu/Speech/Transcript%20-%20IG%20Metall%20 Conference%20December%202012%20(1).pdf, accessed 10 June 2013.
27. *The Guardian*, 18 June 2014.
28. *The Guardian*, 16 April 2014.
29. The origins of this formulation can be found in a radically different political context in Michel Foucault, *The History of Sexuality Volume 2: The Use of Pleasure*, Trans. R. Hurley, (London: Penguin Books, 1984) p. 25ff.
30. *European Convention on Human Rights*, 1 June 2010, p. 5, European Court of Human Rights, Council of Europe, www.echr.coe.int.
31. David Cameron, 25 January 2012, Speech on the European Court of Human Rights, transcript at https://www.gov.uk/government/speeches/speech-on-the-european-court-of-human-rights, accessed 12 February 2014.
32. *European Convention on Human Rights*, 1 June 2010, p. 5.
33. For examples see: http://www.washingtonpost.com/blogs/blogpost/post/ angela-merkel-depicted-as-nazi-in-greece-as-anti-german-sentiment-grows/2012/02/10/gIQASbZP4Q_blog.html, accessed 1 June 2013; http:// www.joe.ie/news/world-affairs/angela-merkel-depicted-as-a-nazi-on-cover-of-greek-newspaper/, accessed 1 June 2013.
34. Volker Kauder, 15 November 2011, http://www.spiegel.de/international/ europe/now-europe-is-speaking-german-merkel-ally-demands-that-britain-contribute-to-eu-success-a-798009.html, accessed 21 May 2013.

35. Angela Merkel, 16 May 2013, http://www.spiegel.de/international/berlin-calls-nazi-comparison-by-hungarian-leader-a-derailment-a-900830.html, accessed 21 May 2013.
36. Angela Merkel, 16 May 2013.
37. *The Times*, 19 February 2014.
38. John Maynard Keynes, *The Economic Consequences of the Peace* (New York: Harcourt, Brace and Howe, 1920) Project Gutenberg E-text by Rick Niles and Jon King, http://www.gutenberg.org/cache/epub/15776/pg15776.txt, accessed 12 May 2013.

10   Epilogue

1. IPCC Working Group III Contribution to AR5, 'Climate Change 2014: Mitigation of Climate Change', http://mitigation2014.org/, accessed 25 April 2014.
2. Unmanned Aerial Vehicle or Remotely Piloted Aircraft.
3. Numerous opinion polls point to support for this statement. See, for example, 'Public Trust in Banking', 2013, YouGov-Cambridge University opinion research, http://cdn.yougov.com/cumulus_uploads/document/ylf7gpof19/Public_Trust_in_Banking_Final.pdf, accessed 25 April 2014.
4. American researchers Martin Gilens and Benjamin Page reach similar conclusions using a contrasting, quantitative methodology. Their findings suggest that American policy making by an affluent elite runs the risk of being more oligarchy than democracy. See Martin Gilens and Benjamin I. Page, 'Testing Theories of American Politics: Elites, Interest Groups, and Average Citizens', Princeton University and Northwestern University, 9 April 2014, http://www.princeton.edu, accessed 14 April 2014.

Bibliography

Adams, J., The Works of John Adams, Vol. 6, *Defence of the Constitutions Vol. III* (Boston: Charles C. Little and James Brown, 1851).

Atlas Collaboration, 'A Particle Consistent with the Higgs Boson Observed with the ATLAS Detector at the Large Hadron Collider', *Science*, Vol. 338 (2012) pp. 1576–1582.

Augustine, *The City of God Against the Pagans* (Cambridge: Cambridge University Press, 1998).

Banks, E., *See No Evil: Uncovering the Truth behind the Financial Crisis* (Basingstoke: Palgrave Macmillan, 2011).

Bartelson, J., *A Genealogy of Sovereignty* (Cambridge: Cambridge University Press, 1995).

BBC Charter, 28 June 2006, 'Regulatory obligations on the UK Public Services', *Broadcasting: An Agreement Between Her Majesty's Secretary of State for Culture, Media and Sport and the British Broadcasting Corporation* (London, The Stationery Office, 2006).

BBC News, http://www.bbc.co.uk/news/

BBC Trust Report, June 2007, 'From Seesaw to Wagon wheel: Safeguarding Impartiality in the 21st Century' p. 40, located at http://www.bbc.co.uk/bbctrust/assets/files/pdf/review_report_research/impartiality_21century/report.pdf, accessed 20 July 2012.

Bell, D. (ed.), *Ethics and World Politics* (Oxford: Oxford University Press, 2010).

Bellamy, A., *Just Wars: From Cicero to Iraq* (Cambridge: Polity Press, 2006).

Benjamin, M., *Drone Warfare: Killing by Remote Control* (New York and London: OR Books, 2012).

Berlin, I., (1958) *Two Concepts of Liberty*. In Isaiah Berlin (ed.) *Four Essays on Liberty* (Oxford: Oxford University Press, 1969).

Blackburn, S. and Simmons, K. (eds), *Truth* (Oxford: Oxford University Press, 2010).

Blair, T., *A Journey*, (London: Hutchinson, 2010).

Bones, C., *The Cult of the Leader: A Manifesto For More Authentic Business* (Chi Chester: John Wiley & Sons Ltd, 2011).

Carlson, M., 'A Brief History of the 1987 Stock Market Crash with a Discussion of the Federal Reserve Response', The Federal Reserve, November 2006, http://www.federalreserve.gov/pubs/feds/2007/200713/200713pap.pdf, accessed 14 March 2014.

Carter, R. M., *Climate: The Counter Consensus* (London: Stacey International, 2010).

Chappelle, W., McDonald, K. and McMillan, K., *Important and Critical Psychological Attributes of USAF MQ-1 Predator and MQ-9 Reaper Pilots According to Subject Matter Experts* (Wright-Patterson AFB, OH: Air Force Research Laboratory, 2011).

Charter of the United Nations, https://www.un.org/en/documents/charter/

Churchill, W., 11 November 1947, Speech to Parliament, Hansard, Vol. 444, cc. 207, http://hansard.millbanksystems.com/commons/1947/nov/11/parliament-bill#column_207, accessed 2 February 2014.

Climate Change Act, 2008, The National Archives, http://www.legislation.gov.uk/ukpga/2008/27/contents, accessed 20 March 2013.

Cole, C., Dobbing, M. and Hailwood, A., *Convenient Killing: Armed Drones and the 'Playstation' Mentality* (Fellowship of Reconciliation: Oxford, 2010).

Connolly, W. E., *The Augustinian Imperative*, 2nd Edition (Oxford: Rowman & Littlefield Publishers, 2002)

Davies, H., *The Financial Crisis: Who is to Blame?* (Cambridge: Polity Press, 2010).

Davis, R., 'The "Problem of de Gaulle": British Reactions to General de Gaulle's Veto of the UK Application to Join the Common Market', *Journal of Contemporary History*, Vol. 32, No. 4 (1997) pp. 453–464.

Delingpole, J., *Watermelons: How Environmentalists are Killing the Planet, Destroying the Economy and Stealing Your Children's Future* (London: Biteback Publishing, 2012).

Dessler, A. and Parson, E. A., *The Science and Politics of Global Climate Change*, 2nd Edition (Cambridge: Cambridge University Press, 2010).

Drake, S., *Galileo at Work* (Chicago: University of Chicago Press, 1978).

Drone Wars UK, http://dronewars.net/

Elshtain, J. B., 'Women and War: Ten Years On', *Review of International Studies*, Vol. 24 (1998) pp. 447–460.

Elshtain, J. B., *Just War Against Terror* (New York: Basic Books, 2004).

Elshtain, J. B., *Sovereignty: God, State, and Self* (New York: Basic Books, 2008).

Ereaut, G. and Segnit, N., *Warm Words: How Are We Telling the Climate Story and Can We Tell It Better?* (London: Institute for Public Policy Research, 2006).

European Convention on Human Rights, 1 June 2010, p. 5, European Court of Human Rights, Council of Europe, www.echr.coe.int

Feyerabend, P., *Against Method*, 3rd Edition (London and New York: Verso, 2008).

Financial Crisis Inquiry Commission, *The Financial Crisis Inquiry Report: Final Report of the National Commission on the Causes of the Financial and Economic Crisis in the United States* (New York: Public Affairs, 2011).

Fortin, E. L. and Kries, D. (eds), *Augustine: Political Writings*, Trans. M. W. Tkacz and D. Kries, (Indianapolis: Hackett, 1994).

Foster, L. K. and Vince, S., *California's Women Veterans: The Challenges and Needs of Those Who Served* (California, California Research Bureau, 2009).

Foucault, M., 'Truth and Power', in Colin Gordon (ed.), *Power/Knowledge* (New York: Pantheon Books, 1980).

Foucault, M., *Power/Knowledge: Selected Interviews and Other Writings by Michel Foucault, 1972–1977* (New Pork: Pantheon Books, 1980).

Foucault, M., *The History of Sexuality, Volume 2: The Use of Pleasure*, Trans. R. Hurley, (London: Penguin Books, 1984).

Foucault, M., 'The Ethics of the Concern for Self as a Practice of Freedom', in Michel Foucault and Paul Rabinow (eds), *The Essential Works of Michel Foucault 1954 – 1984 Volume 1: Ethics – Subjectivity and Truth* (New York: The New Press, 1997).

Foucault, M., *The Hermeneutics of the Subject: Lectures at the Collège de France 1981– 1982*, F. Gros (ed.), Trans. G. Burchell, (New York and Basingstoke: Palgrave Macmillan, 2001).

Foucault, M., *Society Must Be Defended: Lectures at the Collége de France, 1975–1976* (New York: Picador, 2003).

Fukuyama, F., *The End of History and the Last Man* (London and New York: Penguin Books, 1992).

Funtowicz, S. O. and Ravetz, J. R., 'Science for the Post-Normal Age', *Futures*, Vol. 25, No. 7 (1993) pp. 739–755.

Funtowicz, S. O. and Ravetz, J. R., 'Post-Normal Science', *International Society for Ecological Economics* (2003) 1–10, http://leopold.asu.edu/sustainability/sites/default/files/Norton, per cent20Post per cent20Normal per cent20Science, per cent20Funtowicz_1.pdf, accessed 10 December 2013.

Geominne, G., 'Has Science Ever Been Normal? On the Need and Impossibility of a Sustainability Science', *Futures*, No. 43 (2011) pp. 627–636.

'Global-average temperature records', Met Office, http://www.metoffice.gov.uk/climate-guide/science/temp-records, accessed 9 January 2014.

Goldsmith, E. R. D., Editorial, *The Ecologist*, Vol. 1, No. 1 (July 1990) pp. 2–5.

Goldstein, J. S., *War and Gender: How Gender Shapes the War System and Vice Versa* (Cambridge: Cambridge University Press, 2001).

Gordon, C. (ed.), *Power/Knowledge* (New York: Pantheon Books, 1980).

Grantham Research Institute on Climate Change and the Environment, London School of Economics, 'Anthropogenic Global Warming "stopped" in 1997 ... and in 1996, 1995, 1982, 1981, 1980, 1979, 1978 and 1972', http://www.lse.ac.uk/GranthamInstitute/Media/Commentary/2012/February/anthropogenic-global-warming-1997.aspx, accessed 9 January 2014.

http://www.lse.ac.uk/GranthamInstitute/news/anthropogenic-global-warming-stopped-in-1997and-in-1996–1995–1982–1981–1980–1979–1978-and-1972/, accessed 18 July 2014.

Hargie, O., Stapleton, K. and Tourish, D., 'Interpretations of CEO Public Apologies for the Banking Crisis: Attributions of Blame and Avoidance of Responsibility', *Organization*, Vol. 17, No. 6 (November 2010) pp. 721–742.

Heinberg, R., *The Party's Over: Oil, War and the Fate of Industrial Societies* (Gabriola Island, BC: New Society, 2003).

Heinberg, R., *The End of Growth: Adapting to Our New Economic Reality* (Gabriola, Clareview, 2011).

Hindmoor, A. and McConnell, A., 'Why Didn't They See It Coming? Warning Signs, Acceptable Risks and the Global Financial Crisis', *Political Studies*, Vol. 61 (2013) pp. 543–560.

Hood, C., *The Blame Game: Spin, Bureaucracy, and Self-Preservation in Government* (Princeton NJ: Princeton University Press, 2011).

Hulme, M., *Why We Disagree About Climate Change* (Cambridge: Cambridge University Press, 2011).

Human Rights Watch World Report 2013, Libya, http://www.hrw.org/world-report/2013/country-chapters/libya, accessed 10 February 2014.

Hunt, K., '"Embedded Feminism" and the War on Terror', in K. Hunt and K. Rygiel (eds), *(En)Gendering the War on Terror: War Stories and Camouflaged Politics* (Aldershot: Ashgate, 2006) pp. 51–72.

Hunt, K. and Rygiel, K. (eds), *(En)Gendering the War on Terror: War Stories and Camouflaged Politics* (Aldershot: Ashgate, 2006).

International Human Rights and Conflict Resolution Clinic of Stanford Law School (Stanford Clinic) and the Global Justice Clinic at New York University School of Law (NYU Clinic), Report: *Living Under Drones: Death, Injury and*

Trauma to Civilians from US Drone Practices in Pakistan, report located at http://livingunderdrones.org/report/, accessed 7 November 2012.

IPCC, *Climate Change 2007: Synthesis Report. Contribution of Working Groups I, II and III to the Fourth Assessment Report of the Intergovernmental Panel on Climate Change*, Core Writing Team, R. K. Pachauri and A. Reisinger (eds), (Geneva: IPCC, 2007).

IPCC, 20 January 2010, 'Statement on the Melting of Himalayan Glaciers', located at http://www.ipcc.ch/pdf/presentations/himalaya-statement-20january2010.pdf, accessed 1 September 2012.

IPCC Fourth Assessment Report: Climate Change 2007, 10.5.4.6 Synthesis of Projected Global Temperature at Year 2100, located at http://www.ipcc.ch/publications_and_data/ar4/wg1/en/ch10s10-5-4-6.html, accessed 10 December 2013.

IPCC Fourth Assessment Report: Climate Change 2007, The Himalayan Glaciers, 10.6.2. http://www.ipcc.ch/publications_and_data/ar4/wg2/en/ch10s10-6-2.html, accessed 29 December 2013.

Jones, W. K., Smith, J., Kieke, Jr, B. and Wilcox, L., 'Female Genital Mutilation. Female Circumcision. Who is at Risk in the U.S.?' *Public Health Reports*, Vol. 112, No. 5 (Sep–Oct 1997) pp. 368–77.

Kaza, S., *Mindfully Green: A Personal and Spiritual Guide to Whole Earth Thinking* (Boston and London: Shambhala Publications: 2008).

Keynes, J. M., *The Economic Consequences of the Peace* (New York: Harcourt, Brace and Howe, 1920) Project Gutenberg E-text by Rick Niles and Jon King, http://www.gutenberg.org/cache/epub/15776/pg15776.txt, accessed 12 May 2013.

King, A. and Schneider, B., *The First Global Revolution: A Report by the Council of the Club of Rome* (Orient Longman, 1991).

Kuhn, T. S., *The Structure of Scientific Revolutions*, 3rd Edition (Chicago and London: The University of Chicago Press, 1996).

Lasswell, H. D., *Politics: Who Gets What, When, How*, (Michigan: Whittlesey House, 1936).

Lee, P., *A Genealogy of the Ethical Subject in the Just War Tradition*, PhD Thesis, accepted by King's College London, 1 August 2010.

Mancini, G. F., 'Europe: The Case for Statehood', *European Law Journal*, Vol. 4, No. 1 (March 1998) pp. 29–42.

Mann, M. E., *The Hockey Stick and the Climate Wars* (New York and Chichester: Columbia University Press, 2012).

Marshall, B. J. and Warren, J. R., 'Unidentified Curved Bacilli in the Stomach of Patients with Gastritis and Peptic Ulceration', *The Lancet*, Vol. 323, No. 8390 (June 1984) pp. 1311–1315.

Merton, R. K., 1942, 'The Normative Structure of Science', in N. W. Storer (ed.), *The Sociology of Science: Theoretical and Empirical Investigations* (Chicago: University of Chicago Press, 1973).

Micklethwait, J. and Woolridge, A., *God is Back: How the Global Rise of Faith is Changing the World* (London and New York: Penguin Books, 2009).

Minsky, H. P., *Stabilizing an Unstable Economy* (New Haven, Yale University Press, 1986).

National Academy of Sciences, *On Being a Scientist: A Guide to Responsible Conduct in Research*, 3rd Edition (Washington: The National Academies Press, 2009).

Neely, C. J., 'The Federal Reserve Responds to Crises: September 11th Was Not the First', *Federal Reserve Bank of St. Louis Review*, Vol. 86, No. 2 (March/April 2004) pp. 27–42.

Norman, R., *Ethics, Killing and War* (Cambridge: Cambridge University Press, 1995).

Olsthoorn, P., *Military Ethics and Virtues: An Interdisciplinary Approach for the 21st Century* (New York: Routledge, 2011).

Osteen, J., *Your Best Life Now* (New York: Faith Words, 2004).

Otto, J. L. and Webber, Capt. B. J., 'Mental Health Diagnoses and Counselling Among Pilots of Remotely Piloted Aircraft in the United States Air Force, http://timemilitary.files.wordpress.com/2013/04/pages-from-pages-from-msmr_mar_2013_External_causes_of_tbi.pdf accessed 25 January 2014.

Popper, K. R., *Objective Knowledge: An Evolutionary Approach* (Oxford and New York: Oxford University Press, 1979).

Protocol Additional to the Geneva Conventions of 12 August 1949, and relating to the Protection of Victims of International Armed Conflicts (Protocol I), 8 June 1977, http://www.icrc.org, accessed 20 January 2014.

Raina, V. (2009), 'Himalayan Glaciers – A State-of-Art Review of Glacial Studies, Glacial Retreat and Climate Change', Ministry of Environment & Forests, Government of India, http://moef.nic.in/sites/default/files/MoEF per cent20-Discussion per cent20Paper per cent20_him.pdf, accessed 25 April 2013.

RCM, RCN, RCOG, Equality Now, UNITE, *Tackling FGM in the UK: Intercollegiate Recommendations for Identifying, Recording, and Reporting* (London: Royal College of Midwives, 2013).

Reichberg, G. M., Syse, H. and Begby, E. (eds), *The Ethics of War* (Oxford: Blackwell Publishing, 2006).

Rodgers, P., *United States Constitutional Law: An Introduction* (Jefferson, North Carolina: McFarland, 2011).

Roubini, N. and Minh, S., *Crisis Economics: A Crash Course in the Future of Finance* (London and New York: Penguin Books, 2011)

Schiller, R. J., *The Subprime Solution: How Today's Global Financial Crisis Happened, and What to Do About It* (Princeton: Princeton University Press, 2008).

Schneider, S. H., October 1989 Interview with *Discover* Magazine, Reprinted in Detroit News Editorial Response, 5 December 1989, http://stephenschneider.stanford.edu/Publications/PDF_Papers/DetroitNews.pdf, accessed 17 December 2012.

Singer, P. W., *Wired for War* (New York: Penguin, 2009).

Soon, W. and Baliunas, S., 'Proxy Climatic and Environmental Changes of the Past 1000 years', *Climate Research*, Vol. 23 (2003) pp. 89–110.

Storer, N. W. (ed.), *The Sociology of Science: Theoretical and Empirical Investigations* (Chicago: University of Chicago Press, 1973).

Supreme Court Judgement, 19 June 2013, 'Smith, Ellis, Allbutt and Others *vs.* The Ministry of Defence', [2013] UKSC 41, http://www.supremecourt.gov.uk/decided-cases/docs/UKSC_2012_0249_Judgment.pdf, accessed 26 June 2013.

Tarski, A., 'The Semantic Conception of Truth and the Foundations of Semantics', *Philosophy and Phenomenological Research*, 4 (1944), in S. Blackburn and K. Simmons (eds), *Truth* (Oxford: Oxford University Press, 2010).

Transatlantic Trends 2013 Survey, http://trends.gmfus.org/files/2013/09/TTrends-2013-Key-Findings-Report.pdf, accessed 2 February 2014.

Transparency International Corruption Perceptions Index, shttp://www.transparency.org/cpi2013/results, accessed 7 February 2014.

Treaty of Rome, 25 March 1957, http://ec.europa.eu/economy_finance/emu_history/documents/treaties/rometreaty2.pdf, accessed 5 June 2013.

Tregenna, F., 'The Fat Years: The Structure and Profitability of the US Banking Sector in the Pre-Crisis Period', *Cambridge Journal of Economics*, Vol. 33 (2009) pp. 609–632.

Tucker, B., 'Lynndie England, Abu Ghraib, and the New Imperialism', *Canadian Review of American Studies*, Vol. 39, No. 1 (2008) pp. 83–100.

Turner, A., '*The Turner Review: A Regulatory Response to the Global Banking Crisis*', The Financial Services Authority, Publication Ref 003289, March 2009.

UK Government White Paper, June 2011, 'The Natural Choice: Securing the Value of Nature', (London: The Stationery Office, 2011).

United Nations Framework Convention on Climate Change, Kyoto Protocol, http://unfccc.int/kyoto_protocol/items/2830.php, accessed 15 December 2013.

United Nations Human Rights Council, 9 April 2013, 'Report of the Special Rapporteur on Extrajudicial, Summary or Arbitrary Executions, Christof Heyns', http://www.ohchr.org/, accessed 27 January 2014.

United Nations Security Council Report, 13 June 2013, 'The Situation in Afghanistan and Its Implications for International Peace and Security', http://www.securitycouncilreport.org/, accessed 27 January 2014.

United Nations Security Council Resolution 1244, 10 June 1999, 'Security Council Welcoming Yugoslavia's Acceptance of Peace Principles, Authorizes Civil, Security Presence in Kosovo', https://www.un.org/News/Press/docs/1999/19990610.SC6686.html, accessed 10 February 2014.

United Nations Security Council Resolution 1973, 17 March 2011, http://www.un.org/en/ga/search/view_doc.asp?symbol=S/RES/1973(2011), accessed 10 February 2014.

Uniting and Strengthening America by Providing Appropriate Tools Required to Intercept and Obstruct Terrorism (USA Patriot Act) Act of 2001, Public Law 107-56—26 October 2001, http://www.gpo.gov/fdsys/pkg/PLAW-107publ56/pdf/PLAW-107publ56.pdf, accessed 12 December 2013.

Walzer, M., *Just and Unjust Wars*, 4th Edition (New York: Basic Books, 2006).

White, L., 'The Historical Roots of our Ecologic Crisis', *Ecology and Religion in History*, (New York: Harper and Row, 1974), http://www.uvm.edu/~gflomenh/ENV-NGO-PA395/articles/Lynn-White.pdf, accessed 1 January 2014.

Whitmarsh, L., 'What's in a Name? Commonalities and Differences in Public Understanding of "Climate Change" and "Global Warming"', Public Understanding of Science, Vol. 18 (2009) pp. 401–420.

Index

Printed and bound by CPI Group (UK) Ltd, Croydon, CR0 4YY